PROFESSIONALS TALK LOGISTICS
SUSTAINING STRATEGY AND OPERATIONS

PROFESSIONALS TALK LOGISTICS

Sustaining Strategy and Operations

JON KLUG
STEVE LEONARD

Howgate Publishing Limited

Copyright © 2025 Jon Klug and Steve Leonard

First published in 2025 by
Howgate Publishing Limited
Station House
50 North Street
Havant
Hampshire
PO9 1QU
Email: info@howgatepublishing.com
Web: www.howgatepublishing.com

All rights reserved.

No part of this publication may be reproduced, stored in a retrieval system, or transmitted in any form or by any means including photocopying, electronic, mechanical, recording or otherwise, without the prior permission of the rights holders, application for which must be made to the publisher.

British Library Cataloguing-in-Publication Data
A catalogue record for this book is available from the British Library

ISBN 978-1-912440-67-2 (pbk)
ISBN 978-1-912440-68-9 (hbk)
ISBN 978-1-912440-69-6 (ebk - EPUB)

Jon Klug and Steve Leonard have asserted their right under the Copyright, Designs and Patents Act, 1988, to be identified as the author of this work.

The views expressed in this book are those of the author and do not necessarily reflect official policy or position.

Contents

Foreword Retired General Gus Perna, U.S. Army	vi
Introduction JON KLUG AND STEVE LEONARD	1
PART I: PAST: FROM OXEN TO STUDEBAKERS	4
1. Horses, Camels, or Oxen? How the Great Captains Defined the Art of Logistics STEVE LEONARD	5
2. Does Logistics Drive Strategy, or Does Strategy Drive Logistics? JOE WALDEN	19
3. What Ever Happened to the Arsenal of Democracy? TIM GILHOOL AND SYDNEY SMITH	35
4. Strategy Short of War RON GRANIERI	49
5. The Four Logistical Operations RICHARD KILLBLANE	57
PART II: PRESENT: SUSTAINING CONTEMPORARY WAR	86
6. An Armor Officer's Perspective on Logistical Literacy RICH CREED	87
7. The Operational Level of War and Logistics KEVIN BENSON	100
8. At the End of a 6,000-Mile Screwdriver FRANCIS PARK	112
9. Logistics, Operational Warfare, and the War in Ukraine JIM GREER	130
PART III: FUTURE: BRAVE NEW LOGISTICAL WORLD	146
10. Toward a Theory of Supply Chain Environment MATT EVERS	147
11. Artificial Intelligence and Logistics on the Modern Battlefield STACY TOMIC, MICHAEL POSEY, AND PAUL LUSHENKO	169
12. Darwin Strategic Bastion MICK RYAN	186
13. Contributors	195
Index	201

Foreword

Good logistics is tough business. It's very technical. And you have to take the time to learn it, the systems, the processes, the enablers, and the capabilities that help the organization do all of the required jobs. But what is Logistics? Doctrinal terms adopted by various militaries can sometimes confuse the subject. For the United States Army, as defined by Field Manual 4-0, the overarching concept for supporting military activities is Sustainment. This includes "the provision of logistics, financial management, personnel services, and health service support necessary to maintain operations until successful mission completion (ADP 4-0)." Logistics itself involves "maintenance, transportation, supply, field services, distribution, operational contract support (OCS), and general engineering." Financial Management, Personnel Services, and Health Service Support are considered separate functions under the warfighting function of Sustainment. If it strove to be doctrinally correct, this volume could easily be called 'Professionals Study Sustainment.' But while the idea of Sustainment as a Warfighting Function grew from the World War II Army Services Forces, there are few if any rousing, hard-hitting quotes associated with the term.

Thus, the inspiration for the work before you. It has been attributed to World War II era United States Army General Omar Bradley that "Amateurs talk strategy, Professionals talk logistics." Military Logisticians, from the time of antiquity, have had a thankless task. Alexander the Great is recorded as saying that his Logisticians will be 'the first ones that I slay' if an operation goes wrong. American Revolutionary War leader Nathanial Greene, who served as Washington's chief logistician, lamented that the position was bereft of glory and 'no one has ever heard of a Quartermaster in History.' And just as the work of logisticians often goes unrecognized, the historiography of Military Logistics compared to other topics in the study of war and conflict is not as robust as the topic merits. For 'Professional Study Logistics,' Jon Klug and Steve Leonard have assembled a diverse and talented team of authors to discuss the logistical challenges of yesterday, today, and tomorrow.

The range of topics this team of authors cover is impressive. Former U.S. Army School of Advanced Military Studies director Kevin Benson mines his experience as U.S. 3rd Army Chief of Plans during Operation Iraqi Freedom to discuss the Operational Level of War and Logistics. Retired armor officer and current U.S. Army Combined Arms Command Chief of Doctrine Rich Creed gives us his perspective on Logistical Literacy. James Greer, a retired cavalry officer and former U.S. Army School of Advanced Military Studies director, provides keen insight into sustaining Ukraine.

In addition to these great Combat Arms leaders, the team is well-represented by current and former logistics professionals. Marine Corps logistician Matthew Evers dives deep to offer a theory of a supply chain environment. Retired Army logisticians and serving Department of the Army civilians Sydney Smith and Tim Gilhool explore the decline of the Arsenal of Democracy. U.S. Army Transportation Corps Regimental Historian Emeritus Richard Killblane describes the four major logistics operations. Retired Army Logistics Corps Colonel Joe Walden asks and answers the question: Does logistics drive strategy or does strategy drive logistics?

Lastly, there is great representation of some of today's most thoughtful and insightful strategic thinkers. Steve Leonard, a former logistics officer and retired Army strategist, looks into how history's great captains defined the art of logistics. The Army War College instructor team of Stacy Tomic, Michael Posey, and Paul Lushenko investigate the effects of artificial intelligence and drones on the logistics of the wars of tomorrow. Former U.S. Army Basic Strategic Arts Program director Francis Park looks into sustaining Operation Enduring Freedom. All this is capped off by another great and relevant work of fiction by retired Australian Major General Mick Ryan as he tells the story of Establishing the Darwin Strategic Bastion, which is vital for the Second Pacific War.

Together, this formidable team of past and current military professionals have crafted an impressive work highlighting the vast spectrum encompassing logistics and warfare. In putting this together, they have performed a genuine service to the greater defense community on a relevant, meaningful, and far too often misunderstood topic. If I could add a favorite quote of my own on this matter from U.S. Army General George C. Marshall, "Infantry wins battles, but logistics wins wars." May our Nation and its allies always remember that unchanging imperative as we face the challenges of the 21st Century and beyond.

General Gus Perna, USA, Retired

INTRODUCTION

"Amateurs talk strategy. Professionals talk logistics."
— General of the Army Omar Bradley

It is an adage offered around many a military conference room, a time-honored piece of wisdom often attributed to Omar Bradley, one of Dwight D. Eisenhower's field commanders during World War II and later the first Chairman of the U.S. Joint Chiefs of Staff. It is a cautionary reminder that logistics both enables and constrains military strategy and ignoring one imperils the other.

Military strategy is most assuredly informed by logistics, and contemporary operations reveal that leaders require more professional—and nuanced—discussion of military strategy and logistics. British scholar-practitioner Major General J. F. C. Fuller wrote, "Surely one of the strangest things in military history is the almost complete silence upon the problem of supply. Not in ten thousand books written on war is there to be found one on this subject." Although Fuller's comment was somewhat overstated when he wrote it in the 1930s, his point remains valid—few good book-length works have been written specifically on logistics. Classics, such as Martin Van Crevald's 'Supplying War', John Lynn's 'Feeding Mars' and Henry Eccles's 'Logistics in the National Defense', are rare exceptions. The good news is recently more books have been written on logistics, such as Jobie Turner's 'Feeding Victory', which explores five cases across three centuries, and Jeremy Black's 'Logistics', which is a sweeping survey from 1453 to the near future. But more discourse is needed.

The edited volume you are reading—Professionals Talk Logistics—is an effort to continue the recent trend of works on logistics with a focus on the relationship between contemporary military strategy and logistics. The authors and editors hope to add to the discussion of military strategy and logistics by focusing on twentieth- and twenty-first examples. The idea is to convey the significance of military strategy and logistics in recent history and evaluate its current and future practice. Overall, this edited volume argues that for military strategy—often described as ends, ways, means, and risk—logistics remains a means, circumscribes the potential ways, and plays a vital role in determining the time horizon necessary to achieve the desired ends depending on the level of risk.

Professionals Talk Logistics builds on the relatively scant literature on the subject, focusing on the systemic and often complex relationship between military strategy and logistics. Drawing on a wealth of experienced and insightful strategists and logisticians, Professionals Talk Logistics explores that relationship from antiquity into the contemporary context and concludes with a cautionary and provocative view through a directed lens of future conflict.

Part I: Past examines the evolution of strategy and logistics from the campaigns of the great battle captains of antiquity through the rise of the Arsenal of Democracy. In "Horses, Camels, or Oxen?" Steve Leonard explores how the art of logistics is rooted in the great campaigns of the ancient world. Joe Walden's chapter, "Does Logistics Drive Strategy or Does Strategy Drive Logistics?" uses history to answer the age-old question. With, "Whatever Happened to the Arsenal of Democracy?" Tim Gilhool and Sydney Smith discuss the rise of the American defense industrial base while lamenting its current state. Ron Granieri returns readers to the days following World War II to revisit the powerful link between strategy and logistics that played out during the Berlin Airlift. Finally, Richard Killblane closes the opening section with his chapter, "The Four Logistical Operations," which details how history shaped modern military logistics.

Part II: Present explores the relationship between logistics and strategy in a contemporary context. Richard Creed opens the section with "An Armor Officer's Perspective on Logistical Literacy," a personal reflection on the development of logistical knowledge and combat arms expertise over the course of a military career. In "The Operational Level of War and Logistics," Kevin Benson—who led the strategic planning effort for the 2003 invasion of Iraq—provides a fascinating glimpse into the world of strategic formulation, dispelling a number of myths along the way. Francis Park's chapter, "At the End of a 6,000-Mile Screwdriver," offers an informative perspective on how logistics affected operational art during the war in Afghanistan. Jim Greer closes the section with "Logistics, Operational Warfare, and the War in Ukraine," a brilliant exploration of the lessons learned from Ukraine's ongoing war with Russia.

Part III: Future casts a furtive glance forward in time to explore what lies beyond the horizon for logistics and military strategy. In his chapter, "Toward a Theory of the Supply Chain Environment," Matthew

Evers applies a hypothetical perspective to the future of logistics and warfare, examining how various factors may one day shape the complex environment of supply chains. Stacy Tomic, Michael Posey, and Paul Luchenko follow with a perspective on how artificial intelligence already impacts contemporary operations and how that impact may evolve in the future. Finally, bestselling author and media pundit Mick Ryan concludes with "Establishing the Darwin Strategic Bastion," which offers a somewhat haunting view of logistics and strategy in a future second Pacific war.

Throughout history, logistics consistently proved as pivotal during campaigning as the strategy itself. Operational reach, which determines the time and distances over which a campaign may be conducted, is a measure of logistics. The same essentially holds for culmination, which defines the place and time that the combat power is insufficient to achieve the mission, driven largely by when the logistical capacity of a campaigning force has been exhausted. Yet the historical record of the great campaigns is typically overshadowed by the more dramatic—and often heroic—narrative of the captains of battle who led them. As historian Jonathan Roth so aptly noted, "Logistics is least observable when it works well, and usually only enters the historical record when it breaks down."

In the spirit of Omar Bradley's wise observations, the twelve chapters that comprise Professionals Talk Logistics aim to inspire discourse and debate while informing audiences of the vital importance logistics plays in the formulation and execution of military strategy while celebrating the creativity and imagination each author brings to this book. We hope to provide an enlightening experience that helps leaders, strategists, and logisticians better understand both military strategy and logistics and the intrinsic and often complex relationship they share. We hope you enjoy reading it as much as we have enjoyed writing it.

PART I

PAST
FROM OXEN TO
STUDEBAKERS

1

HORSES, CAMELS, OR OXEN?

How the Great Captains Defined
the Art of Logistics

Steve Leonard

On an early spring day in 2002, a small group of United States Army planners gathered in a non-descript sensitive compartmented information facility in an unremarkable government building in St. Louis to sketch the initial strokes for the 2003 invasion of Iraq and the eventual march to Baghdad. It was an eclectic group with varied backgrounds and experiences drawn from division- and corps-level organizations across the Army. Some brought combat experience; others did not. Some were so-called "Jedi Knights," graduates of the School of Advanced Military Studies, the organization that had proven so impactful during Operation Desert Storm. In all, the assembled group represented roughly half of the Army's combat power and included maneuver, intelligence, fires, and logistics planners.

After brief introductions, an officer stood at the front of the room, where a map of the region was projected onto a screen. He gestured with a pointer at the nation of Kuwait, tapping the screen several times. He explained that we would not be executing OPLAN 1003-98, General Anthony Zinni's plan that called for a three-corps attack involving some 380,000 troops. Instead, we would execute a "generated start," a more modest plan that would "compress the deployment phase from sixty to thirty days."[1] In a triumphal tone that evoked George C. Scott's portrayal of General George Patton in the 1970 film *Patton*, the officer proudly stated,

---

1 Micheal Gordon and Bernard Trainor, COBRA II: The Inside Story of the Invasion and Occupation of Iraq (New York: Pantheon Books, 2006), 36-37.

6 Professionals Talk Logistics: Sustaining Strategy and Operations

"We're going to move five divisions into Kuwait in thirty days."[2]

The room buzzed with conversation, with some planners nodding to acknowledge the broad brush stroke that had just been described. From an operational maneuver perspective, it was a bold move that would put ready combat forces on the ground in a relatively short period. When combined with prepositioned assets already in Kuwait, it would put significant combat power on the border before Iraq's vaunted Republican Guard could effectively respond. It was audacious. It was brilliant. It was impossible.

"The math doesn't work," I said from my chair in the third row.

"What do you mean?" the officer responded.

"There isn't enough throughput capacity," I replied, sounding more like a logistician with every word. This was a math problem—one of simple numeracy—with which I was very familiar, having contended with a similar situation during an exercise only a year earlier that required some very creative thinking to overcome the throughput constraints. "You've got one airport and one seaport, and you're competing with civilian traffic at both. You might be able to do some transloading at the container port further to the south, but that has serious depth limitations you'll have to overcome. It can be done, but not in thirty days."

The officer pondered my statement for a moment. He took a breath, then sighed. "We'll take over both the APOD and SPOD.[3] Problem solved."

The room erupted. People argued over one another, and several moments passed before calm was restored.

"That doesn't solve the basic math problem, assuming that Kuwait would even allow us to shut down their air and seaports for a month," I answered. "There simply isn't enough throughput capacity to do it. We might be able to get three divisions through…maybe four with the PREPO stocks at Doha.[4] But not five."

---

2 *Patton*, directed by Franklin J. Schaffner (1970; Burbank, CA: 20th Century Studios). In a pivotal scene in the film, General Dwight D. Eisenhower's senior commanders are gathered to discuss options to relieve the 101st Airborne Division, which was surrounded by German forces in the early hours of the Battle of the Bulge. Patton dramatically declares, "I can attack with three divisions in forty-eight hours."
3 APOD and SPOD are military acronyms for Aerial Port of Debarkation and Seaport of Debarkation, common terms used in deployment planning.
4 Among the Army's global prepositioned stocks, a divisional armored brigade task force is maintained at Camp Doha, Kuwait. Army Prepositioned Stock (APS) – 5 has served as a deterrent in the Middle East since the end of the 1991 Gulf War.

He considered my response for a brief moment before taking a deep breath and responding, "Then we'll launch the fifth division from the north through Turkey."

Another voice joined the rousing debate: "We're going to have to get permission from Turkey. They're not going to just let march through and attack another Muslim country from their territory."

"It won't be a problem," the officer said while gesturing with his pointer. "They're a member of NATO. They'll have to let us through."

They are. They didn't. That last division would take a little longer to deploy into theater, and the generated start would be a little lighter than initially planned.[5]

Emergence and Evolution: The Principles of Sustainment

Throughout the history of warfare, logistics typically assumed a role as pivotal during campaigning as the strategy itself, although the effort necessary to sustain those campaigns was often overshadowed by the more dramatic historical narrative. As historian Jonathan Roth so aptly noted, "Logistics is least observable when it works well, and usually only enters the historical record when it breaks down."[6] While influential battle captains and their heroic exploits captivated researchers through time, logistics consistently proved essential in shaping the outcomes of warfare. Campaigns that ventured beyond one's own borders relied on a decidedly artful combination of local sources and forage to ensure that forces—men, animals, and equipment—could endure the heavy demands placed on them by the rigors of expeditionary warfare. Those great battle captains were successful because they were as adept at the art of sustaining their armies as they were employing them.

Xenophon's account of the Ten Thousand, the retreating Greek mercenaries who fought for Cyrus the Younger in his attempt to seize control of the Achaemenid Empire in 401 BCE, in *Anabasis* remains one of the most insightful studies of logistics operations in ancient warfare.[7] Following the death of Cyrus in the Battle of Cunaxa, the Ten Thousand were stranded deep in enemy territory and forced to mount a fighting

---
5 Gregory Fontenot, E.J. Degan, and David Tohn, *On Point: The United States Army in Operation Iraqi Freedom*. (Leavenworth, KS: Combat Studies Institute Press, 2004), 45-47.
6 Jonathan P. Roth, *The Logistics of the Roman Army at War (264 B.C. – AD 235)*. (Boston: Brill, 1999), 3.
7 Xenophon, *The Anabasis*, trans. John J. Owen. (New York: Leavitt & Allen, 1864), ix-xv.

retreat over two years and 3,100 miles. Xenophon's ability to sustain his force in the heart of a hostile empire over such a long period of time while under near-constant assault was an unprecedented act of military and logistics genius for which historian Theodore Dodge declared him the greatest general prior to Alexander the Great.[8]

Like Xenophon, the son of Philip of Macedonia inherently recognized the logistical challenges presented when campaigning with a sizeable military force. Alexander not only considered logistics in every aspect of his strategic planning, but he also weighed the sustainment needs of his force against changes in weather, terrain, and even harvest calendars.[9] Minute details that might be overlooked by lesser battle captains, such as the different consumption rates for water and feed for various species of pack animal, were key elements of Alexander's planning.[10] The success of Alexander's expeditions was primarily due to his unparalleled grasp of the art of logistics and "his meticulous attention to the provisioning of his army."[11]

The Romans embraced the lessons of Alexander and combined his methods with an established system of depots and magazines and an advanced road network into an incredibly sophisticated yet flexible logistics system that supported—and to a large degree allowed—the empire's expansion. Although the Romans did not have a specific term for logistics, it was important enough to them to consider the provisioning of the Legions a central field of professional study.[12]

As with Alexander, logistics was a central consideration in planning for the Romans. Traditionally, March 1 marked the beginning of campaign season for the Legions, mainly due to the availability of fodder.[13] Once afield, the Romans relied on a well-developed system of supply lines to provision their forces, from traditional overland methods to waterborne transport—both sea and river—as well as through local foraging and requisitioning.[14] And Roman planning doctrine of the time emphasized the necessity of securing these supply lines from enemy attack.[15]

---

8 Theodore Ayrault Dodge, *Great Captains: A Course of Six Lectures*. (New York: Houghton Mifflin, 1890), 7.
9 Donald W. Engels, *Alexander the Great and the Logistics of the Macedonian Army*. (Berkeley, CA: University of California Press, 1978), 119.
10 Engels, 18.
11 Ibid., 3.
12 Vegetius, "The Military Institutions of the Romans," in *Roots of Strategy*, ed. T.R. Phillips. (Mechanicsburg, PA: Stackpole Books, 1940), 128-130.
13 Roth, 279-280.
14 Ibid., 281.
15 Vegetius, 128-130.

Unlike the Romans, the Mongols under Genghis Khan employed a much more austere and adaptive logistics system suited to their way of war. Their exceptional equestrian skills and knowledge of the steppes enabled them to attack with unprecedented speed and mobility. Sustaining such techniques demanded efficient, disciplined logistics planning and organization. While the Mongols relied on supply trains in much the same way as earlier armies, they would abandon their baggage and carts upon entering enemy territory and subsist almost indefinitely on borts (dried meat) and kumis (fermented milk), allowing a hardened Mongol warrior to advance as much as 60 miles in a day. To the Mongols, logistics and strategy were one and the same.[16]

As warfare continued to evolve and armies became vastly larger, increasingly lethal, and exponentially more advanced, the principles of logistics that emerged centuries earlier remained essentially unchanged, even if not explicitly codified. In the same way, the modern principles of war grew from the experiences of the great captains of battle, the observations of military theorists, and the fundamental context of war, the principles of logistics evolved from millennia of sustaining expeditionary campaigns. Those principles—integration, anticipation, responsiveness, simplicity, economy, survivability, continuity, and improvisation—are as immutable as the very nature of war itself.[17] They would be as recognizable to Vegetius or Genghis Khan in their time as they are to military logisticians today.

Integration. Coordinating and synchronizing logistics was central to the success of every great battle captain and was deeply woven into the fabric of their campaign planning. The Romans possessed probably the most complicated logistics system in ancient warfare, and integration and synchronization were essential to their way of war. For Alexander, who campaigned over far greater distances, often operated in remote regions distant from his supply lines, and moved with the minimum amount of provisions necessary, integration was a priority in his planning.

Anticipation. No one considered the provisioning needs of their forces as deeply as Alexander, who—as noted previously—often operated far from his supply lines and relied on his experience and insight to anticipate those

---

16 Wayne E. Lee, "When the Mongols Set Out to Conquer the World, There Was Only One Limiting Factor: Grass," *HistoryNet*, March 15, 2022, https://centurionsreview.com/mongol-logistics/.
17 Department of the Army, *Sustainment*, ADP 4-0. (Washington, DC: Department of the Army, 2019), 1-2-1-4.

requirements and avoid unnecessary pauses or culmination. Alexander's remarkable insight was a product of a prodigious mind honed through study and observation.[18]

Responsiveness. The logistics system that supported the Roman Legions was a model of efficiency and responsiveness, ensuring that their forces afield could consistently acquire what they needed when they needed it. As Roth noted, "A sophisticated logistical system allowed the Romans to exploit their military recourses effectively."[19] Responsiveness ensured the Legions remained provisioned even at the height of battle.

Simplicity. By virtue of their fighting methods, both Xenophon and Genghis Khan depended on uncomplicated sustainment systems suited to their particular circumstances. Fighting what was, in effect, a two-year rearguard action compelled Xenophon to rely on simplicity throughout their march. The Mongols, on the other hand, wielded simplicity as a tool of warfare; their entire logistics system was an outgrowth of their nomadic culture and the Mongolian horse. They could campaign anywhere the steppe grasses grew, and the fattened horses would, in turn, sustain the warriors.[20]

Economy. Like those of Genghis Khan, Alexander's campaigns were models of efficiency. First, Alexander inherited the most effective fighting force in Europe and Asia, which his father, Philip II, had transformed to be based on the Macedonian phalanx.[21] His soldiers had to carry their own equipment and provisions while marching, carts were rarely allowed, and they subsisted on what they needed and nothing more.[22]

Survivability. The sophistication of Roman logistics relied on a distribution system built on supply lines that had to be protected. The survivability of those supply lines—as well as the trains that traversed those lines—was essential, a lesson the Romans learned—and relearned—on numerous occasions.[23] Consuls would fill the ranks of their Legions with their best troops, assigning less-capable garrison soldiers to secure the logistics routes, which proved problematic over time.[24]

Continuity. No better examples of this tenet exist than in the campaigns of Alexander and the Romans. Uninterrupted provisioning was critical to

---

18 Engels, 9.
19 Roth, 279.
20 Li Yulin and Sun Xiaoyan, "Symphony of Nature and Life: Mongolian Horse Culture," *Education About Asia*, Vol. 18, No. 3 (2013), 17-20.
21 Engels, 12.
22 Ibid.
23 Roth, 281-293.
24 Ibid. 285.

High Contact– High Movement	High Contact– Low Movement	Low Contact– High Movement	Low Contact– Low Movement
Mobile Warfare	Methodical Warfare	Maneuver Warfare	Entrapment Warfare
Roving Warfare		Positional Warfare	

Table 1.1　Framework for Warfare Matrix

maintaining the speed and momentum of a campaign while ensuring that the army did not culminate unexpectedly due to logistics failure. Both recognized this fundamental principle and integrated it into their strategic planning from the outset.

Improvisation. Every great battle captain understood the necessity of improvisation, but no one practiced it quite as well as the Mongols. In times of great duress, when all means of sustainment had been exhausted, a Mongol horseman would cut into their mount's neck and drink a small amount of blood for nutrition before applying herbal medication to the wound. This practice was harmless to a healthy mount and provided life-saving sustenance to the horseman.[25]

However, this list of principles is incomplete. If there is perhaps one missing, it would be *numeracy*. This principle is the ability not just to demonstrate cognitive dexterity with numbers, but to interpret the narrative those numbers represent and understand their influence on campaigning. That elevates numeracy to a form of art. Alexander is the exemplar of this principle, and the lengths to which he studied and planned for his army's sustainment requirements represent an unparalleled skill in data analytics.[26] Numeracy allowed Alexander to see beyond the numbers, capture the narrative within the data, and make informed strategic decisions based on that narrative.

The Cognitive Advantage: Strategic Thinking

Although these fundamental principles trace their roots to antiquity and were expanded upon in the writings of Clausewitz, Jomini, and others, they are often relegated to academic discourse—much like the contemporary principles of war—and typically considered only in

---

25　Yulin and Xiaoyan, 19.
26　Engels, 2-3.

retrospective analysis. In educational environments, they are likely to be committed to a mnemonic, memorized, quizzed, and forgotten. In practice, they are routinely neglected: nowhere in a formal military planning process will you encounter such principles. Whether an omission resulting from oversight or a flawed assumption, the fact remains. Yet, for the great captains of battle, the principles were inherent to the intricate tapestry of warfare, so ingrained in their thinking that their application became second nature. They rose to become great captains not because of their conquests—which were remarkable in their own right—but because they possessed a natural ability to think and act strategically that transcended their adversaries. They became great captains because they could assess a complex and evolving situation, recognize and interpret unique patterns, and make critical decisions exponentially faster than anyone else.[27] They could see what others could not, understood their opponents better than they understood themselves, and translated their cognitive prowess into a decisive competitive advantage.

That ability was cultivated at an early age. In some, education was a prominent factor; in others, their natural gifts were focused through the lens of experience. In 343 BCE, when Alexander was just 13 years old, his father brought Aristotle to Pella to serve as a tutor to the future king.[28] When Philip died in 336 BCE and Alexander ascended to rule Macedon, the master returned to Athens, and the pupil set out to conquer the known world. Genghis Khan, on the other hand, had no formal education. As Dana Pittard wrote in a 1986 *Army Review* article, "He never read a book, was never a student of any warlord, was never tutored by scholars."[29] Instead, Khan gained his mastery of strategic thinking through practical experience and forged an empire that spread across much of Asia and Eastern Europe. But, regardless of how that ability was honed, it was sharpened to a fine edge and applied on the field of battle, leading vast campaigns that seem unfathomable in retrospect.

In contemporary terms, strategic thinking exists at the intersection of three closely related fields of study: systems thinking, cognitive psychology, and game theory.[30] Even in antiquity, the mental processes

---

27 Aaron K. Olson and B. Keith Simerson, *Leading with Strategic Thinking* (New York: Wiley, 2015), xiv-xvi.
28 Felix C. Robb, "Aristotle and Education," *Peabody Journal of Education* 20, no. 4 (1943), 204.
29 Dana J. H. Pittard, "Genghis Khan and the 13th-Century AirLand Battle," *Military Review*, Vol. 66, No. 7 (1986), 19.
30 Olson and Simerson, 1.

that underpin these fields had a profound influence on how the great captains planned and prosecuted their campaigns. In warfare, success has always been firmly rooted in strategic thinking. This truism was reinforced by Aaron Olsen and B. Keigh Simerson, whose expertise in leading major organizations is captured in their book, *Leading with Strategic Thinking*. As they noted, "While the focus and process of strategic thinking may vary, it is always key to maintaining perspective and focus when the external environment changes."[31]

Systems thinking—understanding how the external environment behaves and how its various elements interact with and influence one another—represents a level of cognition that was likely as rare then as it is today. Warfare presents inherently complex challenges: multifaceted, interrelated, and in a constant state of change.[32] Cause and effect share a distinctly nonlinear relationship; feedback and adaptive behavior within the external environment combine to create a form of dynamic complexity that defines the system itself.[33] Strategically, the link between ends, ways, and means is equally dynamic, a reality that is as difficult to contend with today as it was during the march of Xehophon's Ten Thousand.

As an academic field of study, cognitive psychology—the exploration of human perception and the mental process that defines how people solve problems, make decisions, and become motivated—is relatively young but, in retrospect, has always been a strength of great military and political leaders. Cognitive psychology reveals "how preconceived notions and beliefs influence and impact your analysis, the conclusions you draw, and the decisions you make."[34] For the great captains, this would have manifested as a measure of how they perceived the external environment, interpreted existing and emerging opportunities, and envisioned a desired future state.

Finally, game theory—the study of decision-making in a competitive, dynamic, and multifaceted system—examines the complex interplay between risk, opportunity, and change when pursuing that desired future state. "Rather than assuming a neutral or positive reaction from stakeholders, game theory expects and anticipates other parties' potential disagreement,

---

31 Ibid., xxvi.
32 Barry Richmond, "Systems Thinking: Critical Thinking Skills for the 1990s and Beyond," *System Dynamics Review*, Vol. 9, No. 2 (1993), 113-33.
33 Peter M. Senge, *The Fifth Discipline: The Art and Practice of the Learning Organization*. (New York: Doubleday, 1990), 71.
34 Olson and Simerson, 3-5.

conflict, and action that may hinder progress.[35] That competitive mindset fueled the success of the great captains, who recognized and accepted risk as a catalyst rather than something to be avoided. As a result, they were able to create opportunities again and again, driving their campaigns forward.

Ironically, perhaps, each of these fields is a separate area of scientific study that can only be fused into a cohesive whole through the application of strategic art. With the great captains, their significant cognitive abilities allowed them to not only master the constant ebb and flow of the complex adaptive system that represented their external environment, but to extend that mastery in time and space. By applying the artistic component so fluidly, they could execute lengthy campaigns across vast distances and sustain them. Where lesser captains found logistics to be a limiting factor in their conquests, the great captains prevailed.

Napoleon's Glance: Strategic Intuition and Phronesis

What differentiated the great captains from their contemporaries was their ability to make qualitatively better decisions faster: faster than their opponents, faster than the situation evolved, faster than circumstances might normally allow. They instinctively embraced risk. They accepted uncertainty as a natural element of warfare. They possessed highly developed and disciplined intuitive abilities that fueled their decision-making.

Intuition has long been considered an essential leadership attribute, a key trait that helps to speed decision-making processes and reduce reaction time in times of duress. Among accomplished strategic leaders, intuition is a pivotal skill that allows them to receive, interpret, and act on information instinctively, without any conscious reasoning. Like any other principal leadership attribute, it is forged through study, practice, and reflection. And when honed with wisdom and judgment, it becomes a powerful strategic weapon.

Just as neuroscience informs the understanding of strategic thinking, it has influenced how intuition is perceived, developed, and applied. In broad terms, leaders use three levels of intuition: basic, expert, and strategic.[36] Basic or ordinary intuition is closely associated with the instinctual feeling that drives a response. Basic intuition is purely emotional; it involves no

---

35 Ibid., 16-18.
36 Stephanie Vozza, "How to Use Your Intuition to Get Ahead," *Fast Company*, March 4, 2014, https://www.fastcompany.com/3027157/how-to-use-your-intuition-to-get-ahead.

discernable thought and is driven by the brain's limbic system. Simply put, it is a 'sense' that urges action. Instinct is a 'fight or flight' feeling; high-risk decisions should not depend on basic intuition.

Expert intuition, on the other hand, is rooted in pattern analysis, "a form of rapid thinking where you jump to a conclusion when you recognize something familiar."[37] What Malcolm Gladwell described as recognition-primed decision-making—in which rapid-fire decisions are possible through the leverage of well-developed mental models—was, in truth, expert intuition.[38] Expert intuition is the hallmark of many successful leaders, and that innate sense is honed as mental models mature and develop. It fuels increased self-confidence, comfort with ambiguity and uncertainty, and willingness to take risks. It drives courage of conviction, personal assertiveness, and a willingness—even an eagerness—to draw conclusions and make decisions in the moment. Mental models—which fuse pattern recognition and decision-making—form the basis of much of the research of psychologist Gary Klein, whose pioneering work in decision-making is underpinned by a deep study of expert intuition.[39]

Unlike basic or expert intuition, strategic intuition is rooted in logic and reason rather than emotion. It is, according to Duggan, "thinking, not feeling."[40] When facing a new situation where patterns are not immediately recognizable, the brain requires additional time to make the neural connections necessary to provide reasoned insight. "The discipline of strategic intuition requires you to recognize when a situation is new and turn off your expert intuition. You must disconnect the old dots, to let new ones connect on their own."[41]

In essence, strategic intuition was what Clausewitz described when he referred to *coup d'œil*: "the inward eye…the quick recognition of a truth that the mind would ordinarily miss or would perceive only after long study and reflection."[42] As Dr Antulio Echevarria notes, "*coup d'œil*

---

37 William Duggan, *Strategic Intuition: The Creative Spack in Human Achievement*. (New York: Colombia Business School, 2007), 1-2.
38 Malcolm Gladwell, *Blink: The Power of Thinking Without Thinking*. (New York: Little, Brown and Company, 2005).
39 Gary Klein, Sources of Power: How People Make Decisions. (Boston, MA: MIT Press, 1998).
40 Duggan, 2.
41 Ibid.
42 Carl von Clausewitz, *On War*, trans. Peter Paret and Michael Howard. (Princeton, NJ: Princeton University Press, 1976), 102.

describes the ability to see simultaneously with the physical as well as the mind's eye."[43] It is an apt description considering Clausewitz was referring to one of history's great battle captains, Napoleon Bonaparte. But "Napoleon's glance" also offers a cautionary tale. Duggan warns that "expert intuition can be the enemy of strategic intuition."[44] As leaders gain experience and insight, they begin recognizing patterns in similar circumstances, developing the mental models that allow them to make rapid decisions and solve problems faster. "Strategic intuition," Duggan explains, "requires [that] you recognize when a situation is new and turn off your expert intuition."[45]

The situation Duggan describes is the intuition trap. When a new situation presents itself or circumstances are unfamiliar, leaders must possess the humility and discipline to allow for strategic intuition to take root. But as often is the case, hubris and ego—the overwhelming need to make a decision in the heat of the moment—can lead to making poor decisions based on flawed or imperfect mental models or a misinterpretation of events. Whether due to an insatiable zeal for glory or simply a flawed instinct for battle, Napoleon's glance, the strategic intuition that brought his greatest victories, also brought his greatest defeats.

Alexander, conversely, presents a somewhat contrasting example of strategic intuition. While his campaigns have been the subject of intense study for over two millennia, his application of strategic intuition in sustaining those campaigns goes largely unnoticed.[46] In his book, *Alexander the Great and the Logistics of the Macedonian Army*, historian Donald W. Engels describes Alexander's innate ability to leverage systems thinking when considering the provisioning of his army, his uncanny logistics genius, and his intuitive sense of how best to conduct a resupply to avoid disrupting operational momentum.[47] Whether provisioning by sea or by land, carrying supplies with the march columns or storing them in predesignated locations, or relying on local sources of forage and supply, Alexander's "glance" consistently served him well and allowed him to extend his long campaigns across distances that would still prove logistically challenging today.

---

43 Antulio J. Echevarria II, *Clausewitz and Contemporary War*. (Oxford: Oxford University Press, 2007), 109.
44 Duggan, 2.
45 Ibid.
46 Engels, 1-4.
47 Ibid., 2-3.

What differentiated Alexander's glance from that of Napoleon may have been phronesis, the Aristotelian concept of an elevated form of practical wisdom that fused experience, judgment, knowledge, and reflection.[48] Although Aristotle first wrote of phronesis in Book VI of *Nicomachean Ethics* in 350 BCE, 14 years after leaving Macedon for Athens, there is little doubt the subject—one of the three forms of knowledge professed by Aristotle—was a point of emphasis during his tutoring of Alexander.[49] Phronesis may have been the Greek's greatest lesson imparted to the young Macedonian.

Among Aristotle's classifications for knowledge—including *episteme* (scientific knowledge) and *techne* (knowledge of craft)—phronesis implied an application of judgment and reflection not necessarily present in the others.[50] Throughout his leadership of the Academy, Aristotle maintained the value of phronesis, which the poet Samuel Coleridge referred to as "common sense to an uncommon degree."[51] Such common sense, even among the great captains, was—in a word—uncommon. And nowhere was it more important than in sustaining their campaigns.

The Data Whisperers: The True Art of Logistics

Interpreting and drawing predictive insights from the external environment is a practiced skill that requires years to master; making sound, anticipatory decisions based on those insights is a form of art largely dependent on phronesis. The same holds true with data. As a professional field, logistics tends to focus on quantitative skills: interpreting order-ship times, calculating stockage levels and reorder points, coordinating maintenance schedules, and managing distribution. In the same vein, nearly every aspect of maintaining readiness—from personnel strength to training to operational readiness rates—has roots in numeracy.

While the principles of sustainment lean more toward the science of logistics, numeracy—especially concerning anticipatory decision-making—requires bridging the quantitative and the qualitative. In contemporary debate on the subject, the discourse typically centers on whether data is

---

48 Peter Massinghan, "An Aristotelian Interpretation of Practical Wisdom," *Palgrave Communications*, Vol. 5, No. 123 (2019). https://doi.org/10.1057/s41599-019-0331-9.
49 Aristotle, *Nicomachean Ethics*, trans. W.D. Ross. (London: Oxford University Press, 1925), 142-144.
50 Ibid.
51 Ikujiro Nonaka and Hirotaka Takeuchi, "The Big Ide: The Wise Leader," *Harvard Business Review* (May 2011), 4.

a commodity or whether decision-making is rooted in data analytics. In reality, the only argument should be whether data-centricity can be elevated to a form of art.

Even in antiquity, data-centric or data-informed decision-making was not uncommon. Engels's study of Alexander's campaigns reveals a practiced use of detailed logistics data that transcends anything seen again until the advent of modern warfare.[52] The operational techniques employed by the Mongols suggest an unparalleled precision to their planning with respect to logistics.[53] And the complexity of the Roman logistics system reflects a similar focus on sustainment data.[54] What set those efforts apart from others was not the use of data because the data alone only told a part of the story.

Capturing the full context required a practiced eye and a sharp mind to understand the narrative within the numbers and an experienced hand to make informed decisions based on that narrative. It required phronesis. The great captains succeeded by interpreting and applying that data in predictive analysis. They sustained their long campaigns by listening to the narrative, by leveraging phronesis. Where others might see the numbers and make decisions based on raw data, the great captains were data whisperers. They listened closely to the narrative within the numbers and made informed, predictive decisions in what we would call "real-time" today. In his own way, in his own time, Aristotle was the original data whisperer, a skill he inspired in Alexander and one shared by the greatest of the great battle captains. While historians through the ages paid scant attention to the mundane details of sustaining campaigns that often spanned years, those details were what made the great captains true masters of the art of war.

---

52 Engels, 11-25.
53 Yulin and Xiaoyan, 19.
54 Roth, 279.

2

DOES LOGISTICS DRIVE STRATEGY OR DOES STRATEGY DRIVE LOGISTICS?

Joe Walden

The question this chapter's title poses is not an idle one, as it reminds several communities of the need to craft military strategy and logistical planning holistically. In an attempt to answer the larger question, it offers a historical look at logistics driving strategy and strategy driving logistics. While looking at these historical examples of strategy driving logistics and logistics driving strategy, it is essential at all levels to keep in mind the guidance that General Fred Franks provided the logisticians at Camp Arifjan, Kuwait, before the initiation of Operation Iraqi Freedom: "Forget logistics and you lose." As a definition of military strategy, the United States Marine Corps manual, *Warfighting*, states that "strategy can be thought of as the art of winning wars."[1] Taking Franks' adage a bit further and combining it with *Warfighting*'s point, logistics are required to win wars; therefore, strategy and logistics must be intertwined.

We first need to define logistics to explore which influences which. Why is it essential to define logistics? It is important to set the foundation for what logistics is, and professional organizations' definitions provide several examples. From a military perspective, it is supply, transportation, movement of personnel and equipment, distribution, and maintenance. However, the commercial world looks at logistics a little differently. For example, the Association for Supply Chain Management (ASCM) defines logistics as "the subset of supply chain management that controls the

---

1   United States Marine Corps. Marine Corps Doctrinal Publication (MCDP) 1, *Warfighting*. (Washington, D.C.: U.S. Government, April 4, 2018), 28.

forward and reverse movement, handling, and storage of goods."[2] This chapter explores the influences of logistics on strategy and strategy on logistics in historical context from both military and commercial perspectives, attempting to determine if there is an answer to the question: "Does logistics drive strategy, or does strategy drive logistics?" However, there may not be a definitive answer, or at least not a one-size-fits-all answer to the question. Ultimately, the answer may be, "It depends."

John Huston began his seminal military logistics work, *The Sinews of War*, with the thought, "Logistics is a subject which few people, including professional soldiers, have thoroughly understood."[3] Everyone seems to think they understand logistics and often misuse the word in everyday conversations. The same is true for strategy, as 'experts' on all major networks profess to understand strategy yet have trouble explaining what it is, much less how it relates to logistics. From a commercial perspective, ASCM defines strategy as identifying "how the company will function in its environment. The strategy defines how to satisfy customers, how to grow the business, how to compete in its environment how to manage the organization and develop capabilities within the business."[4] From a military perspective, "Military strategy is a set of ideas implemented by military organizations to pursue desired strategic goals."[5] British historian and military theorist B.H. Liddel Hart links strategy and logistics in his seminal work, *Strategy*. He describes strategy: "It may be expressed scientifically by saying that, while the strength of an opposing force or country lies outwardly in its numbers and resources, these are fundamentally dependent upon stability of control, morale, and supply."[6]

In this look at historical examples from the military and commercial industry, it would be remiss not to focus on the strategist Sun Tzu.[7] In the first chapter of *The Art of War*, Sun Tzu clearly shows that a link exists between strategy and logistics. He details the way, or the strategy, and later states that to be successful in any endeavor, you must "carefully guard your line of supplies."[8] Several translations of *The Art of War* also credit

---

2 Pittman, Paul H. and Atwater, J. Brian, eds. *ASCM Supply Chain Dictionary*. (Chicago: Association of Supply Chain Mananement, 2022).
3 Huston, John. *Sinews of War*. (Washington, DC : Center for Military History, 1965), v.
4 Pittman and Atwater.
5 MCDP 1, *Warfighting*, 28.
6 Hart, B.H. Liddell. *Strategy*. (Middlesex, England: Meridian Press, 1954), 5.
7 Sun Tzu, *The Art of War*, trans. Sawyer, Ralph D. (New York: Basic Books, 1994).
8 Sun Tzu, *The Art of War*, trans. Arcturus (London: Arcturus Publishing, 2008), 90.

Sun Tzu with saying that logistics is the line between order and chaos.[9] His guidance has survived centuries of military and commercial operations, as we will see in this historical analysis of the link between strategy and logistics.

The next important observations on the link between strategy and logistics come just a few hundred years later with Alexander the Great. Reportedly, Alexander said: "My logisticians are a humorless lot. They know they are the first ones I will slay if an operation fails."[10] Alexander the Great used a form of just-in-time logistics to drive his strategy as he moved across Europe, Asia Minor, and Asia Major. In 323 BCE, Alexander realized that his strategy of maneuver was impacted by logistical constraints, and that his logistics supply trains were slowing his movement. This realization also revealed that the number of wagons and carts needed to feed the horses and oxen was larger than the number of wagons and carts required to transport his soldiers. This led him to adopt an early form of just-in-time logistics: Alexander had his troops eat the oxen and use the wagons as wood to fuel fires. He rethought his strategy to forage for food by capturing areas and taking what he needed for his movements while leaving behind forces to protect the rear and keep the indigenous people happy enough to prevent retaliation. Alexander had to consider the availability of supplies as part of the strategy of movement as he moved across Europe and Asia. From this perspective, supply and logistics were the basis of his strategy and the resultant tactics.

Historian Donald Engels conducted the most comprehensive study of Alexander and the logistics of his army in *Alexander the Great and the Logistics of the Macedonia Army*. Engels looked in detail at the impacts of the availability, acquisition, transport, and distribution of supplies—the foundation of logistics—on the strategy of movement and conquest. Without modern-day technology and intelligence-gathering systems, Alexander's logisticians had to determine the availability of materials and the methodologies for acquiring and distributing essential aspects of supply. There were no simple solutions to logistics. As a result of this, Alexander's strategy was influenced and driven by logistics availability. This situation led to one of the first examples of just-in-time logistics, as his

---

9 Sun Tzu, *The Art of War*, trans. Sonshi. Accessed July 4, 2024: https://www.sonshi.com/sun-tzu-art-of-war-translation-original.html.
10 The Logistics of Logistics, "My Logisticians are a Humorless Lot." *The Logistics of Logistics*. Accessed May 21, 2024: https://www.thelogisticsoflogistics.com/my-logisticians-are-a-humorless-lot/#iLightbox[gallery4085]/0.

army needed to be resupplied approximately every ten days with food and even more frequently with water. The need for just-in-time logistics driven by transport constraints affected the strategy.[11]

Alexander organized his army around fast movement and swift-striking ability. This same concept drove the design of the modern-day light infantry division in the U.S. Army. However, a weakness of this light organization, much like the constraints of light infantry divisions, was the inability to sustain mass transport of logistics supplies. The army's ability to transport approximately ten days of logistics supplies further impacted the strategy for movement. Alexander's approach was further shaped by the same considerations that affect today's strategy and tactics: the climate of the operational area, its geography, and the availability of ports to receive supplies and logistics. This need to be close to ports or river transport was evident in the movements through Greece and Turkey.

Another logistical consideration that impacted Alexander's strategy was the availability of fresh water and the ability to transport sufficient quantities to support the army. The distances between cities and the harvest dates of the regions in question further exacerbated this challenge. Careful logistical planning was needed to ensure sufficient food was available based on harvest dates. Strategic planning dictated arriving in the regions during or immediately after harvest to ensure adequate food and water. Finally, the level of friendliness or hostility of the indigenous peoples affected the strategic movement. Hostile actions dictated not only the availability of food and logistics support but also the security of the food. Much like the strategic planning for recent operations in Iraq and Afghanistan, Alexander's strategists and logisticians learned operations and logistics are challenging in that environment. This is another area where logistics impacted strategy. Engels summarized his study of Alexander by stating that "supply was indeed the basis of Alexander's strategy."[12]

Moving two thousand years forward in time to 1777, it was logistics, or the lack thereof, that circumscribed the strategy of the Continental Army, where no one really understood the complexities of logistics support. In her book *Supplying Washington's Army*, historian Erna Risch states: "Colonial leaders were handicapped by a lack of practical experience with supply agencies, but they were well aware of the importance of both men and

---

11 Engels, Donald W. *Alexander the Great and the Logistics of the Macedonian Army*. (Berkeley, CA: University of California Press, 1978), 28-29.
12 Ibid., 119.

supplies in military operations."[13] The lack of logistics supplies drove the strategy and the operations of the Continental Army. The shortage of logistics capability resulted in the creation of the Quartermaster Corps and the Quartermaster General. These efforts did not solve the problem, but they did make strategic planning for logistics a priority. The well-documented winter encampment at Valley Forge suffered from the lack of logistics, as was the anecdotal comment, "The supply efforts of the Continental Congress have generally been dismissed as inept."[14]

Several decades later, the United States erupted into civil war. The impact of logistics in shaping the strategy of both sides would lead to the conclusion that, at least in this war, logistics drove strategy. The primacy of logistics can be seen in the following examples.

The Civil War was the first war that saw extensive use of rail to move personnel, equipment, and supplies. The ability to move large masses of personnel quickly by rail affected warfare. Soldiers and equipment could now be quickly moved to impact the strategies of both belligerents. The real impact of logistics on strategy using rail can be seen in the U.S. Army's advantage due to the standardization of the rail system in the North and the centralization of control of the rails compared to the multiple rail gauges used in the South and the lack of a centralized office to control rail movements. This lack of standardization in the South impacted the strategic movement. It resulted in delays and, in some cases, the inability to use rail as a combat multiplier or strategic element.[15]

The Union strategy of the blockade of Confederate ports severely affected logistics capabilities and had a resultant impact on the strategy employed by the Confederates. The Confederate plan was to use "blockade runners," or fast ships with dark paint and low profiles, which could slip through the blockade to bring supplies into the Confederate coastal ports. While not always a successful strategy, it did add to the logistics and strategy link during the war.[16]

Much like today, in the 1860s the Mississippi River was a vast logistical corridor. The strategy of the Union and Confederate forces was to control this logistics force multiplier. When the U.S. Navy took control of

---

13 Risch, Erna. *Supplying Washington's Army.* (Washington, DC: Center for Military History, 1981), 6.
14 Ibid., 10.
15 Gable, Christopher R. *Rails to Oblivion: The Decline of the Confederate Railroads in the Civil War.* (Fort Leavenworth, KS: Combat Studies Institute, 2002).
16 O'Harrow, Robert Jr. *The Quartermaster: Montgomery C. Meigs, Lincoln's General, Master Builder of the Union Army.* (New York: Simon & Schuster, 2016).

the mouth of the Mississippi at New Orleans, this impacted the logistical capability of the Confederates. The seizure of Vicksburg in July 1863 ensured Union control of the Mississippi River while at the same time dividing the Confederate States by closing the vital logistical corridor.[17]

One of the first examples of strategy and logistics in the Civil War came in early 1862. The first recipients of the Medal of Honor were volunteers on an operation to sever the strategic logistical lines of the Confederacy. Tasked to capture a Confederate train and subsequently destroy the critical railroad bridges between Atlanta and Chattanooga, the volunteers infiltrated deep behind Confederate lines and met up at Marietta, Georgia. At the time, Atlanta was the largest supply depot of the Confederacy, linking a rail network that ran from the coastal ports on the Atlantic to most of the Southern states. At the time of the operation, Chattanooga held the largest concentration of Confederate forces. By severing the logistical and operational hubs of the Confederacy, Union leaders believed they could bring a quicker close to the war. The strategy was sound, but the operation was a failure: The Great Locomotive Chase, as it came to be known, ended without accomplishing its main objectives and resulted in the execution or imprisonment of most of the volunteers.[18]

The strategy of invading the North while bypassing Washington, D.C., was driven by the need to procure logistical support in the way of food from the local countryside. In a strategy reminiscent of Alexander—moving through fertile areas to procure food and supplies—Robert E. Lee's army marched into Pennsylvania in desperate need of supplies. Once again, this is an example of a sound strategy driven by logistics; a failure of execution resulted in what was most likely the turning point of the war. When coupled with the simultaneous loss of control of the Mississippi River as a key logistical corridor, the tide turned in favor of the Union. However, the North's strategy of attacking the Southern logistics capabilities did not end there.

Once again, controlling the rail networks came into play with the concentration of forces to conquer Atlanta. During the 1860s, Atlanta was the major rail network in the southern states. This rail concentration inevitably led to Atlanta becoming a vast logistics hub for the support of the Confederate Army. Unlike the previous attempts to constrain Confederate

---

17 Ibid.
18 Bonds, Russell S. *Stealing the General: The Great Locomotive Chase and the First Medal of Honor.* (Yardley, Pennsylvania: Westholme Publishing, 2006).

logistics by controlling or destroying the rail lines projecting from the Atlanta hub, Major General William T. Sherman's strategy was to cripple the Confederacy's ability to conduct logistical operations by seizing control of the rail hub itself. Sherman's "March to the Sea" built on a campaign strategy of total war, destroying everything between Atlanta and Savannah, including railroad operations and other logistics support capabilities. The initial strategy was to remove the rails, leaving the lines inoperable. When the Confederates replaced the rails, Sherman brought in special machines to heat the rail tracks up and then wrapped the rails around trees to prevent them from being used again. Here again, logistics drove strategy.[19]

One final example of logistics driving strategy in the American Civil War is the last 'real' battle: the Siege of Petersburg. Logistics drove the strategy that resulted in the siege, as Petersburg was Virginia's last remaining rail hub. General Ulysses S. Grant—who led the Union victory earlier at Vicksburg that secured Union control of the Mississippi—possessed a deep understanding of logistics and knew that the loss of Confederate rail capability at Petersburg should bring a close to the war. His assumption was accurate, as the war was over within a week of the fall of Petersburg. Without logistics capabilities and support, the Confederate Army was forced to surrender.[20]

Logistics continued to grow in importance in the strategic planning of the 20th Century. This can be seen in the American effort to build the Panama Canal. The US fomented a revolution in Panama to secede from Columba and become a separate country, and logistics completely drove the American strategic interest. When the Panama Canal locks opened in 1914, the width of the largest U.S. Navy vessel at the time determined the width of the locks. The strategic interest was the ability to move ships from one ocean to another quickly. Any commercial benefits from this strategy were corollary benefits. Recognizing the logistical impacts on strategy and military movement and the logistical effects on commerce, Japan and Germany developed plans to destroy the Panama Canal during World War II.

The strategic and logistical importance of the Panama Canal remains today. While not driven solely by military strategy, it is the force behind Panama's build-up of logistics clusters. With the strategy to move

---

19 American Civil War Museum, display.
20 American Battlefield Trust. 10 Facts: Appomattox Court House. Accessed May 13, 2024: https://www.battlefields.org/learn/articles/10-facts-appomattox-court-house.

commercial products as rapidly as possible, location analysis comes into play. A quick look at the number of companies that have established operations close to the canal is an excellent example of logistics strategy driving commercial strategy. Free Trade Zones, two large commercial airports, and three major ports provide the capability to shape a commercial strategy to bring products closer to Central and South America and the Caribbean nations. With the opening of new and larger locks in 2016, the strategy of where to ship products became a new issue. Moving the new Post-Panamax ships through the canal instead of unloading larger ships at ports and cross-loading to the rails or smaller vessels to transit the isthmus allowed the enlargement of East Coast Ports in the United States.[21]

World War II is another major conflict that could lead to answering the strategy/logistics question. The war started because of supply and logistics issues. Japan needed supplies and logistics support for its advancement through the Asia-Pacific region. The limits and then restrictions placed on the Japanese trade by the U.S. and other allies prompted the Japanese attack on the forces at Pearl Harbor. Later, the American island-hopping campaign strategy in the Pacific is another example of the need for logistics support bases driving strategy. The need for logistical support shaped the potential invasion of Japan, as the need to have logistics security coupled with the need to have strategic bases.[22]

The planning for the invasion of Europe is another example of logistics driving strategy. "By the end of May 1943 there probably was no staff officer in Washington who was not convinced that logistical considerations were very important factors."[23] This lack of a focused strategy strained the logistics planners and support structure. This lack of focus led to back-and-forth discussions that resulted in, once again, strategy impacting logistics planning. However, the continued refocus of strategy also meant that there was a logistics constraint to support strategic plans in two distinct theaters. So, the result was also logistics driving strategy.

Caught up in this cycle of strategy driving logistics and logistics driving strategy was the planning for the invasion of Europe. Where should the logistical priority lie? The initial strategy to move from Sicily to Italy

---

21 Panama Canal Authority, Panama Canal Celebrates Eighth Expansion Anniversary with New Draft and Daily Transits Increase. Accessed June 28, 2024: https://pancanal.com/en/celebrates-eighth-expansion-anniversary/.
22 Hotta, Eri. Japan 1941: Countdown to Infamy. (New York: Vintage, 2013).
23 Coakley, Robert W., and Leighton, Richard. Global Logistics and Strategy 1943-1945. (Washington, DC: Center for Military History, 1989) 91.

drove the logistics constraint of landing craft availability. This was soon overpowered by the strategy to invade Europe via France. Once again, strategy drove logistics planning, but logistics capabilities drove strategic reality. The reality was that there was not enough logistics capability to support the two proposed amphibious operations in Italy and France. Port capacities, lift capacity, and available air and sea craft for movement of personnel and supplies constrained the deployment planning for Operation Overlord.[24] Like the planning leading up to Operations Desert Shield/Desert Storm almost 50 years later, the strategic plan consisted of types of units needed but not specific units available. This created logistics constraints in moving the proper units to England in time for the cross-channel operations. The study of Operation Overlord linking strategy and logistics could be a complete volume of its own.

As all of this was being debated among the strategists and logisticians, the campaign strategy leading to Operation Market Garden provides another example of strategy driving logistics and logistics driving strategy. In what became known to some American historians as "Montgomery's Folly," a reference to the massive airborne and ground assault into the Netherlands proposed by British Field Marshal Bernard Montgomery, many logistical considerations were contemplated and, in some cases, ignored. The first was a conscious decision to delay capturing and opening the Port of Antwerp to relieve the logistical burden of bringing supplies and equipment into Europe via Normandy. Montgomery's decision to delay opening the Port of Antwerp delayed the logistics capability and constrained critical throughput capacity. Instead, he focused his efforts on the operations that he was certain could end the war early. Operation Market Garden, was unsuccessful. The strategic planning that prompted this action was a great example of flawed strategy driving logistics.[25]

The supply services could not keep up with the rate of attack after the breakthrough at Normandy—like the requirement to have an operational pause in the early days of Operation Iraqi Freedom—another example of logistics driving strategy. According to Charles B. MacDonald, the issue in 1944 was not the ability to build up logistic support; instead, the constraint was the inability to transport the logistical supplies.[26] Once again, logistics drove strategy and the resultant operations. The need to control and open

---

24 Ibid., 233.
25 MacDonald, Charles B. "The Decision to Launch Opeation Market-Garden," in The Siegfried Line Campaign. (Washington, D.C.: Center for Military History, 1954), 429-442.
26 Ibid., 429-442.

ports closer to the areas of operations dictated the need to capture German-controlled ports such as Antwerp. This challenge underscores the need to not only take the Port of Antwerp but also take the Scheldt tributaries leading into Antwerp. General Eisenhower acknowledged the need to clear the supplies out of the Normandy area and the implications on strategic plans and logistical support capabilities if there was bad weather between the time of the Normandy supply build-up and the start of operations in Northern Europe. However, Eisenhower still decided to delay the capture of Antwerp, leading to a complete circle of strategy driving logistics—the need to capture the ports to logistics capabilities driving strategy.[27]

Vietnam Logistics and Strategy

Geography played an essential role in both the First and Second Indochina Wars. For the Indochina War, French occupation logistics drove strategy as bases needed to be near shorelines, ports, or rivers. The defeat of the French at Dien Bien Phu is an excellent example of logistics impacting strategy. The French found themselves too far from inbound ports for logistics resupply when the road networks and the rivers were cut off by Giap's forces. This situation prompted the strategy first to fight and then to flee the area in defeat.[28]

During the Second Indochina War, the Ho Chi Minh Trail (HCMT) logistics corridor drove the strategy of both the Communist Vietnamese and the coalition of the United States and South Vietnamese. For the North Vietnamese and the North Vietnamese Army, the HCMT served as the logistics corridor south and, as a result, helped shape their strategy. However, the need to expand operations reversed this situation, and the strategy drove the need to expand the HMCT. This strategy-driven expansion of the HCMT led to a movement away from porters carrying supplies to using trucks to move supplies, which in turn expanded the logistical demands and planning.[29]

For the U.S., the strategy to control the North Vietnamese Army (NVA) logistics led to defoliation using Agent Orange and Operation Commando Lava, an attempt to soften the soil around the HCMT to prolong the effects of the monsoon season on the trafficability of the trail. The expansion of

---
27 Ibid., 429-442.
28 Nash, N. S. Logistics in the Vietnam Wars. (South Yorkshire: Pen and Sword, 2020), 167.
29 Ibid.

U.S. involvement in Vietnam also meant a new technique of employing helicopters for transportation. Since the Army was decades from adopting a strategy of a single fuel on the battlefield for land and aviation assets, this drove the logistics need to acquire, store, and dispense larger quantities of aviation-grade fuel.

The U.S., not unlike the French before them, entered the conflict with a strategy of overpowering the NVA with logistics power and superiority. This strategy was sound on the surface, as the same strategy worked in the American Civil War; however, much like the initial resistance of the Confederate Forces in the American Civil War, this strategy overlooked the NVA strategy to defend their homeland. Unlike the Confederate Forces in the Civil War, the NVA could win a war of attrition. The NVA strategy to counter the American logistical superiority was using tunnels to supplement the HCMT and for protected logistics storage. These tunnels were throughout South Vietnam and dangerously close to the headquarters in Saigon. One of the largest tunnel complexes for moving personnel and supplies for the NVA was directly beneath the living quarters and home base operations for the 25th Infantry Division. The strategy to provide this living area comprised of "suitable living quarters"[30] Cu Chi drove logistics operations that employed approximately 240 trucks daily over an extended period to move supplies and building materials from Long Bihn to Cu Chi.[31] This use of tunnels helped drive the NVA insurgency strategy. This same use of tunnels helped drive the logistics strategy for the operations of Walt Disney World, as all of the resupply operations for Disney World are under the tourist level to provide unseen logistics support, although for entirely different reasons.

The U.S. also entered the conflict with the idea of fighting an enemy that employed both insurgent forces and conventional forces with a strategy that reflected a conventional war with infantry and armored vehicles, not unlike the initial plans for Operation Iraqi Freedom. However, while supportable logistically, this strategy drove logistics and led to logistical surpluses. As the Communist Vietnamese modified their tactics to meet this strategy, they fulfilled the axiom of Sun Tzu: "He who can modify his tactics in relation to his opponent and thereby succeed in winning may be called a heaven-born captain."[32]

---

30 Ibid.
31 Ibid., 168.
32 Sun Tzu, trans. Arcturus, 62.

When the Army moved to a search-and-destroy strategy, it also developed a chemical defoliant to support this effort. This requirement led to billions of dollars in research and development for what was initially named Agent Purple, now known as Agent Orange.[33] By 1966, American Forces had dispersed over 200,000 gallons of Agent Orange as part of this strategy. This colossal volume exceeded the forecast for dispersal, and once again, strategy execution impacted logistics because of a shortage of products available. This shortage led to logistics, specifically the lack of supplies, affecting strategy. As the Military and commercial industries have learned repeatedly, a strategy that the logistics system cannot support will not be successful. The logistics shortage, budget cuts, and political pressure led to a new strategy of sending all the remaining Agent Orange back to storage sites.[34] This new strategy led to reverse logistics and warehousing impacts or strategy now driving logistics. Further political pressure led to a distribution strategy that used local companies. By most accounts, this strategy led to increased corruption, reduced logistics effectiveness, and reduced logistics capabilities. This strategy was coupled with the new strategy of Vietnamization—once again, strategy drove logistics.[35]

By 1973, the American strategy had become one of withdrawing forces from Vietnam, and a reverse logistics strategy was required. In this case, what was needed was a plan for what would remain in Vietnam and what would retrograde back to the United States. Unfortunately, this plan never materialized, just like at the end of WWII when over 77,000,000 square feet of storage space was filled with materials excess to the needs of the U.S. Army.[36] This same lesson or observation of needing a reverse logistics strategy would surface again after Operation Desert Shield/Desert Storm, Operation Iraqi Freedom, and Operation Enduring Freedom in Afghanistan.[37]

---

33 Nash, 171.
34 Ibid., 177-178.
35 Ibid., 222-227.
36 Walden, Joseph. *Introduction to Operations Management - A real world approach.* (Lawrence, Kansas: University of Kansas Press, 2020), 440.
37 Shamberger, Jason. "Mission Complete: Iraq and Afghanistan," *Defense One*, July 8, 2022. https://www.dla.mil/About-DLA/News/News-Article-View/Article/3085936/mission-complete-iraq-and-afghanistan/.

Operations Desert Shield/Desert Storm

In August 1990, Iraq invaded Kuwait and threatened Saudi Arabia. The U.S. Armed Forces rapidly started to move into Saudi Arabia in a defensive posture. Offensive operations were not possible initially as a result of logistics constraints. Once again, logistics, or the lack thereof, impacted strategy. The U.S. military adopted a push strategy for logistics supplies and materials for the next five and half months. The build-up of logistics support and materials enabled the strategy of freeing Kuwait and removing all Iraqi forces from the country while preventing the ability to reinvade Kuwait.[38] Additionally, General Gus Pagonis' book, *Moving Mountains*, detailed the logistics build-up to support the stated strategy of removing the Iraqis from Kuwait.[39] The build-up strategy was successful, as were the operations that the logistics mountains supported. The execution of the strategy was so successful that the operations only lasted for about four days.

The logistics reality after the offensive operations concluded was that approximately 27,500 twenty-foot equivalent containers were sitting on the docks of the Saudi port of Ad-Dammam. Much like the reality of the reverse logistics after World War II, this logistics build-up resulted from pushing supplies into theater, which required a new logistics strategy to get all the supplies, equipment, and personnel out of the theater and back to home stations. A logistics push strategy works on the philosophy that the shipper knows better what the customer needs than does the customer. This approach is not always wrong for military operations, but it has a history of producing excess materials that must be retrograded and simply sent back after operations.[40]

While the impacts of a logistics push strategy can be seen in almost every operation since World War II, the reality was that the size of the reverse logistics operations increased with each new operation—a new wrinkle in the question of whether strategy impacts logistics or logistics impacts strategy. For example, the movement into Somalia produced excess products so quickly that materials were returned to Germany within days of being shipped. The push strategy produced excess products within days of shipping supplies to Bosnia and Croatia and within hours of starting the humanitarian mission into Uganda and Rwanda.[41]

---

38 Schwarzkopf, Norman, and Petre, Peter. It Doesn't Take a Hero. (New York: Bantam, 1993).
39 Pagonis, William G. Moving Mountains. (Boston: Harvard Business Press, 1992).
40 Peltz, Eric, and Lackey, Arthur. Value Recovery from the Reverse Logistics Pipeline. (Santa Monica: RAND Corporation, 2003).
41 21st TAACOM, *After-Action Review, Uganda/Rwanda*. (Kaiserslautern, Germany: United States Army Europe, 1995).

Operations Iraqi Freedom/Enduring Freedom

It was these memories of logistics excesses produced by the push strategy in previous conflicts that produced strategic and operational consternation during the logistics build-up for Operation Iraqi Freedom. Better strategic planning produced a more feasible Time Phased Force Deployment List. This better strategic planning enabled better logistics planning and execution as well as reduced initial excess materials. The flight path between the units prepping for operations into Iraq and the planning headquarters in Doha, Kuwait, was directly over the Theater Distribution Center, allowing senior leaders and planners a clear view of the daily build-up in the distribution center. This situation produced regular conversations between the director of distribution and the chief planning officer. The concern was to ensure that the excess build-up experienced in Desert Shield/Desert Storm was not repeated. Additionally, much like the logistically created delays experienced by Patton's Third Army in the move across Europe due to logistics constraints, the U.S. Army and Marines experienced a similar delay as they quickly moved across Iraq, resulting in the logistics system's inability to keep up with the rapid movement. While not an example of logistics driving strategy, it is another example of logisticians and strategists not fully communicating.

Commercial (Examples of logistics driving strategy and strategy driving logistics)

The Military does not have a lock on strategy and logistics linkages. The In-N-Out Burger chain is an excellent example of logistics driving strategy in the commercial world. The chain requires delivering fresh daily to stores. This requirement underpins their expansion strategy as they will only expand as far as they can deliver fresh daily to the stores, so the logistics capabilities drive the expansion strategy.[42]

The 2024 Suez Canal and Red Sea issues also provide excellent examples of logistics driving strategy. The attacks and threatened attacks on commercial shipping by the Houthi rebels have created a new logistics strategy of diverting shipments around the Horn of Africa to avoid potential threats. This dictates a new delivery strategy that requires an additional 10,000

---

42 Perman, Stacy. *In-N-Out Burger: A Behind-the-Counter Look at the Fast-Food Chain That Breaks All the Rules*, (New York: Collins Business, 2009).

miles of transit and requires companies to rethink their corporate delivery strategies to meet customer expectations and promised delivery times while lengthening the supply lines.[43]

The environmental impacts of 2023 and 2024 in Panama provide another example of logistics driving strategy. The reduction in rainfall in Panama in 2023 and 2024 lowered Gatun Lake's water levels, supporting not only the freshwater for Panama but also the water required to fill the locks for shipping transit through the Canal. This issue reduced the size of the ships allowed to transit the Canal and limited the number of ships that could pass through the Canal. Thus, the environment changed the shipping and delivery strategies of many companies. The options became shipping to the canal and transloading via rail to the other side for certain commodities and containerized cargo, shipping around the Cape of Magellan, or shipping via air.[44]

The same linkage between logistics and strategy can be seen in clustering logistics activities around the Panama Canal and its associated Free Trade Zones. However, it can also be seen in the clustering of logistics activities around the FedEx hub in Memphis and the UPS World Port in Louisville. The strategy of UPS and FedEx to deliver the next day, second day, or later has driven companies to locate manufacturing and distribution operations in close proximity to the major hubs to allow quick delivery to the hubs to ensure next-day delivery of products.[45]

Amazon and Walmart provide examples of the symbiotic relationship between logistics and strategy. In 1999, Amazon decided to expand away from just selling books and incorporate distribution operations into its strategy. The thought process at the time, well before the strategy of same day/next day delivery, was the need for six distribution centers in the U.S., one in Europe, and one in Asia. Today, Amazon is approaching 400 facilities as they link the strategy of the next day with the need for more distribution centers to support the customer. Walmart has also linked its customer support strategy to its logistics strategy. Walmart came to the realization, much like Target did in the last several years, that each store is a logistics support center in itself. This realization enables Walmart and now Target to ship from stores and not just distribution centers. With over 5000 Walmart stores and over 1900 Target stores, this strategy enables both chains to ship

---

43 Dominguez, Antonio. Interview by Joseph Walden. "President, Latin America and the Caribbean Region." *Maersk*, January 2024.
44 Ibid.
45 Sheffi, Yossi. *Logistics Clusters*. (Cambridge, MA: MIT Press, 2012).

from more locations than the Amazon 300+ distribution facilities in the U.S., possibly becoming more responsive to the customer in the battle for quick delivery.

Conclusion

What does all of this mean from current events and history? The question is: Does logistics drive strategy, or does strategy drive logistics? Military historical examples from 512 BC to 2022 (Sun Tzu to Afghanistan) provide examples of logistics driving strategy and strategy driving logistics. Examples from commercial industry over the past several decades reveal the same trend. In the service sector with the In-N-Out Burger chain and from the distribution sector with Walmart, Target, and Amazon and Sheffi's details of logistics clusters there are current commercial examples of logistics driving strategy and strategy driving logistics.

The goal of this chapter was to determine which one drives the other. The answer through time is that it really depends. It would be easy to develop a narrow focus and state that logistics drives strategy if you are a logistician. At the same time, it would be just as easy for a strategist to support the claim that strategy drives logistics. There are even examples from current events in operations and supply chain management that would allow companies to support either claim.

The bottom line is that neither strategy nor logistics should be planned without considering the other. The two are inseparable. To succeed in any endeavor in the Military or commercial business, strategists and logisticians need to be in sync. Otherwise, operations will go awry, and strategies will fail. There is a reason why after the development of strategies that lead to the development of operations plans, the logisticians then conduct a Rehearsal of Concept drill to walk through the plans to ensure that logistics and strategies are in sync.

Returning to the guidance of General Fred Franks to the logisticians and strategy developers at Camp Arifjan, Kuwait, in late February 2003, "Forget logistics and you lose." This is true for strategists in commercial industry or the Military. At the same time, logistics plans that are not in sync with strategies are useless. The symbiotic relationship between strategy and logistics will continue to provide examples of logistics driving strategy and strategy driving logistics.

3

WHATEVER HAPPENED TO THE ARSENAL OF DEMOCRACY?

Tim Gilhool and Sydney Smith

In December 2004, when being questioned by Army reservists serving in Iraq about the lack of key supplies and other war materials, then American Secretary of Defense Donald Rumsfeld infamously answered, "As you know, you go to war with the army you have, not the army you might want or wish to have at a later time."[1] This bit of unwashed introspection applies equally, if not more so, to the United States Defense Industrial Base. This chapter is an attempt to show how strategic and operational decisions made over the course of decades have impacted the Nation's ability to field and sustain operational forces at home and abroad. It is also meant to be a cautionary tale as the United States tries to best organize and equip for the challenges of Large-Scale Combat Operations (LSCO) in the coming decades of the 21st Century.

How Did We Get Here?

Less than a century ago, there was a massive, interconnected military and commercial infrastructure that supplied the United States and its Allies in the World War II, referred to by then President Franklin Delano Roosevelt as the "Arsenal of Democracy." Images of massed landing ships and mountains of supplies at the Normandy beaches in the aftermath of the Allied invasion in June 1944 encapsulated the industrial and martial might of the United States. However, building that capacity was not a simple affair or even a straight line. Examining what the Nation had to do to get to that point and what it did afterward during the Cold War and beyond

---

1 Kaplan, Fred. "Rumsfeld vs. the American Soldier." Slate Magazine, December 8, 2004. https://slate.com/news-and-politics/2004/12/rumsfeld-vs-the-american-soldier.html.

can help explain how the U.S. got to its present circumstances. It will also hopefully give some insight into what would be required to create such capacity again.

There is one particular industrialist and corporation, along with the city where they dwelled, who would play an important role in shaping what would become the Arsenal of Democracy. While Detroit-based Henry Ford and his Ford Motor Company may not have invented the automobile, he did implement a new way of manufacturing many vehicles quickly—the moving assembly line. The most common feature of this assembly line was the Ford Motor Conveyer belt. The belts had previously been used in other industries, such as slaughterhouses. After much trial and error, in 1913 Ford and his employees successfully began using this innovation at their Highland Park, Michigan assembly plant. The new process allowed the Model T to be built in only 90 minutes. Ford's example and techniques would quickly spread to other automakers and a diverse set of industries.[2] Together with a large, well trained work force centered around the Detroit Metropolitan area, it was an exceptionally solid foundation for future defense industrial production. While Ford is far from alone of corporate examples such as this, it is illustrative of a greater truth: just before the start of the First World War, the United States enjoyed a sizeable industrial base and one of the largest economies in the world. Despite being in a depression at the start of the war, it was well positioned to provide significant material support to the Allied Powers, regardless of its initial neutrality status.[3]

In addition to the automobile industry, another American corporation would play a leading role in setting conditions for the future Arsenal of Democracy. That corporation would dominate the Allied supply chain for things that go boom. Founded in 1802 as the EI du Pont de Nemours and Company by French exiles, the massive munitions company would become the principal supplier of gunpowder to the United States government throughout the 19th Century and deep into the next.[4] The DuPont Corporation would make a fortune during the First World War by supplying the European Allies and later the U.S. Army with high-powered explosives for artillery shells, manufacturing up to 40 percent of

---

2 "The Moving Assembly Line and the Five-Dollar Workday." Ford Corporate. Accessed July 31, 2024. https://corporate.ford.com/articles/history/moving-assembly-line.html.
3 "The Economics of World War I." The National Bureau of Economic Research. Accessed July 31, 2024. https://www.nber.org/digest/jan05/economics-world-war-i.
4 Burclaff, Natalie. "Du Pont: From French Exiles to the Toast of the Brandywine: Inside Adams." The Library of Congress, July 26, 2021. https://blogs.loc.gov/inside_adams/2021/07/du-pont/.

the munitions used by the Allies during the war. DuPont's revenues from the sale of powder and explosives soared from US$25 million in 1914 to US$319 million by 1918, totaling an astonishing US$1.245 billion in this five-year period.[5] These massive profits, which in today's dollars adjusting for inflation is more than US$27 billion,[6] would draw increased public scrutiny, especially from progressive politicians in the years after the First World War. Some were keen to brand the DuPont family and their corporation as "Merchants of Death," although a 1934 book of the same name actually provided a more even-handed and well-researched study of the various military industries and their impact on society.[7]

The Bureaucracies of War and Peace

While American defense industries were rapidly evolving over the course of the early 20th century and establishing deep commercial ties with various European powers, the United States government under the President Woodrow Wilson administration held to a more isolationist foreign policy. Despite this, circumstances eventually drew the United States into the war as a belligerent. What followed after the declaration of war by the U.S. Congress on April 4, 1917, was the establishment of multiple government administrative entities that would attempt to regulate important war-related industries to harness the nation's production capability best. The most prominent of these was the War Industries Board (WIB), which existed from July 1917 to December 1918 to coordinate and channel production in the United States by setting priorities, fixing prices, and standardizing products to support the war efforts of the United States and its allies. Operating for some time under with multiple different leaders and with limited actual enforcement powers, the WIB did not meet the expectations of the Wilson administration. Ultimately, the WIB failed to coordinate the economy to the degree anticipated. It often had to negotiate with industry rather than direct it, and its chairman, Bernard M. Baruch, was never an industrial czar. Other related entities like the National War Labor Board,

---

5 Sass, Erik. "World War I Centennial: Breaking up Dupont." Mental Floss, June 13, 2012. https://www.mentalfloss.com/article/30916/world-war-i-centennial-breaking-dupont.
6 "What is $1.25 in 1918 worth today? inflation calculator for $1.25 since 1918." Accessed August 5,2024. https://www.saving.org/inflation/inflation.php?amount= 1.245+billion+&year=1918&toYear=2024
7 Lindsay, James M., Kat Duffy, and Miles Kahler. "Merchants of Death." Foreign Affairs, January 25, 2024. https://www.foreignaffairs.com/reviews/capsule-review/1934-07-01/merchants-death.

the Emergency Fleet Corporation, and the Federal Railroad Administration (personally run by the Secretary of the Treasury) were established and had varying degrees of success.[8] These entities dissolved with the end of the war, but the experience would help inform subsequent efforts.

In the aftermath of the First World War and the Treaty of Versailles, the United States returned to a general policy of isolationism. After the Stock Market Crash of 1929 and subsequent global economic depression that followed, in 1932 the United States elected a new president with a significant mandate for change. Franklin Delano Roosevelt (FDR) promised to give the American people a 'New Deal,' which began to take shape immediately after his inauguration in March 1933. Based on the belief that the power of the federal government was necessary to get the country out of the depression, the first days of Roosevelt's administration saw the passage of banking reform laws, emergency relief programs, work relief programs, and agricultural programs.[9]

Later, a second New Deal was to evolve; it included union protection programs, the Social Security Act, and programs to aid tenant farmers and migrant workers. Many New Deal acts or agencies came to be known by their acronyms. For example, the Works Progress Administration was the WPA, while the Civilian Conservation Corps was the CCC. While FDR would encounter significant opposition to many of his policies from American industry and his Republican opponents in Congress, these programs would become the foundation to help establish the large administrative state that would enable the soon-in-coming Arsenal of Democracy.[10]

Priming the Pump of Military Industry

By the mid-1930s, events in Europe and Asia indicated that a new world war might soon erupt. To avoid participation in another external conflict, the U.S. Congress took action to enforce a continued policy of U.S. neutrality. On August 31, 1935, Congress passed the first Neutrality Act prohibiting the export of "arms, ammunition, and implements of war" from the United

---

8 "War Industries Board / 1.0 / Encyclopedic - 1914-1918-Online (WW1) Encyclopedia." 1914, July 9, 2024. https://encyclopedia.1914-1918-online.net/article/war-industries-board/.
9 "President Franklin Delano Roosevelt and the New Deal: Great Depression and World War II, 1929-1945: U.S. History Primary Source Timeline: Classroom Materials at the Library of Congress: Library of Congress." The Library of Congress. Accessed July 31, 2024. https://www.loc.gov/classroom-materials/united-states-history-primary-source-timeline/great-depression-and-world-war-ii-1929-1945/franklin-delano-roosevelt-and-the-new-deal/.
10 Ibid.

States to foreign nations at war and requiring arms manufacturers in the United States to apply for an export license. On February 29, 1936, Congress renewed the Act until May 1937 and prohibited Americans from extending loans to belligerent nations.

The Neutrality Act of 1937 did contain one significant concession: belligerent nations were allowed, at the discretion of the President, to acquire any items except arms from the United States, so long as they immediately paid for such items and carried them on non-American ships—the so-called "cash-and-carry" provision. Since vital raw materials such as oil were not considered "implements of war," the "cash-and-carry" clause would be quite valuable to whatever nation could use it. Roosevelt had engineered its inclusion as a deliberate way to assist Great Britain and France in any war against the Axis Powers since he realized they were the only two countries with hard currency and ships to use "cash-and-carry." Unlike the rest of the act, which was permanent, this provision was set to expire after two years.

After a fierce debate in November 1939, Congress passed the final Neutrality Act. This act lifted the arms embargo and put all trade with belligerent nations under the terms of "cash-and-carry." The ban on loans remained in effect, and American ships were barred from transporting goods directly to belligerent ports. However, what these terms and conditions did was to help energize American industry to start producing large amounts of warfighting material over two years before the United States' formal entry into the World War II. In October of 1941, after the United States had committed itself to aid the Allies through Lend-Lease, Roosevelt gradually sought to repeal certain portions of the Neutrality Act. On October 17, 1941, the House of Representatives revoked Section VI, which forbade the arming of U.S. merchant ships, by a wide margin. Following a series of deadly U-boat attacks against the United States Navy and merchant ships, the Senate passed another bill in November that also repealed legislation banning American ships from entering belligerent ports or "combat zones."[11]

---

11 "Milestones in the History of U.S. Foreign Relations – The Neutrality Acts, 1930s" U.S. Department of State. Accessed July 31, 2024. https://history.state.gov/milestones/1921-1936/neutrality-acts.

An Appeal to the Nation

Nazi Germany began their war of conquest with the invasion of Poland in September 1939, followed by offensives in Norway, the Mediterranean, and eventually France, Belgium, and the Netherlands in the summer of 1940. Later, on December 29, 1940, during one of his "Fireside Chats" radio broadcasts to the American people, FDR made his case for establishing the national infrastructure to fight this new global conflict. FDR stated, "We must be the great arsenal of democracy. For us, this is an emergency as serious as war itself. We must apply ourselves to our task with the same resolution, the same sense of urgency, the same spirit of patriotism and sacrifice as we would show were we at war."[12] This clarion call further galvanized American industry to move towards mass production of the ships, planes, tanks, and other weapons of war that would be necessary for victory.

In the aftermath of the Pearl Harbor attack on December 7, 1941, FDR would further set audacious goals for wartime production. His figures shocked Congress, industry, and the nation:

> *"Our goals have been set: This year 60,000 planes—next year 125,000 planes. This year 45,000 tanks—next year 75,000 tanks. This year 20,000 antiaircraft guns—next year 85,000 antiaircraft guns. This year 8,000,000 tons of merchant shipping—next year 10,000,000 tons of merchant shipping... no other nation in the world has ever undertaken or could ever undertake such a program. In 1942 alone we will produce nearly three times as many weapons and supplies of war as in all the eighteen months since the fall of France. In 1942 alone, we will produce as many tanks and planes as Hitler did in all the years before 1939 when he was preparing for world conquest."*[13]

Despite the staggering nature of his request, both Congress and industry would soon answer it with vigor.

---

[12] "Franklin D. Roosevelt - Fireside Chat # 9 - December 29th, 1940." Fireside Chat. | The American Presidency Project, December 29, 1940. https://www.presidency.ucsb.edu/documents/fireside-chat-9.

[13] "Franklin D. Roosevelt - State of the Union Address – January 6th, 1942." State of the Union Address | The American Presidency Project, January 6, 1942. https://www.presidency.ucsb.edu/documents/state-the-union-address-1

Building the Arsenal of 'Bureaucracy'

Perhaps the most essential element of building the Arsenal of Democracy was establishing the administrative structure that would guide and regulate American industry during the war. The War Production Board (WPB) was established in January 1942 with Executive Order 9024.[14] Its primary task was converting civilian industry to war production. The WPB assigned priorities and allocated scarce materials such as steel, aluminum, and rubber, prohibited nonessential industrial production, controlled wages and prices, and mobilized the people through propaganda. The national WPB constituted the Chair, the Secretaries of War, Navy, and Agriculture, an Army Lieutenant General in charge of War Department procurement, the Director of the Office of Price Administration, the Federal Loan Administrator, the Chair of the Board of Economic Warfare, and the Special Assistant to the President for the defense aid program.[15]

The WPB also had broad advisory, policy-making, and progress-reporting divisions. The WPB also employed mathematicians to construct and maintain multilevel models of resources needed for the war effort. Their data models tracked manufacturing defects, materials lost when ships were sunk at sea, and other factors. The WPB managed 12 regional offices and operated 120 field offices nationwide. They worked alongside state war production boards, which maintained records on state war production facilities and helped state businesses obtain war contracts and loans.[16]

Industry Takes Off

With FDR's admonition and the reality of fighting a global war setting in, American industry began to answer. One of the most pressing needs for the rapidly mobilizing United States Army was armored fighting vehicles. Before 1940, the United States did not have a tank production program.

---

[14] "Executive Order 9024-Establishing the War Production Board in the Executive Office of the President and Defining Its Functions and Duties." Executive Order 9024-Establishing the War Production Board in the Executive Office of the President and Defining Its Functions and Duties | The American Presidency Project, January 16, 1942. https://www.presidency.ucsb.edu/documents/executive-order-9024-establishing-the-war-production-board-the-executive-office-the.

[15] "During WWII, Industries Transitioned from Peacetime to Wartime Production." U.S. Department of Defense. Accessed July 31, 2024. https://www.defense.gov/News/Feature-Stories/story/article/2128446/during-wwii-industries-transitioned-from-peacetime-to-wartime-production/.

[16] Ibid.

With the fall of France, American industry began to convert, and after December 1941, transitioned to full-time production of armored fighting vehicles. In response to this need, the Detroit Tank Arsenal Plant sprang up seemingly overnight in the winter of 1940-41 on 113 acres of farmland located north of Detroit in what is now the city of Warren, Michigan. The mammoth structure measured five city blocks deep and two wide. Owned by the government and run by Chrysler, the plant received its first contract to build 1,000 M3 tanks in 1940. The government accepted the first M3 Lee/Grant medium tanks on April 24, 1941, which were built while the plant was still under construction. Beginning in 1942, the plant shifted to build M4 Sherman medium tanks. The plant set an all-time monthly production record by delivering 896 M4s in December 1942. Over the course of the war, the Detroit Arsenal Tank Plant built a quarter of the 89,568 tanks produced in the U.S. In addition to tanks, the U.S. also built over 10,000 M10/M36 tank destroyers, 4,000 M7 Priest self-propelled howitzers, and over 2,000 M26 Pershing heavy tanks.[17]

The American aircraft industry was able to adapt to the demands of war. In 1939 contracts assumed single-shift production, but as the number of trained workers increased, the factories moved to two- and then three-shift schedules. The ultimate example was the Willow Run bomber plant near Ypsilanti, Michigan. It was constructed in 1941 by the Ford Motor Company to mass-produce the B-24 Liberator bomber. The U.S. government contributed $200 million to the project. Built on 975 acres of farmland owned by Henry Ford, the Ford Motor Company developed the site into what was described as the largest war factory in the world. The famous Charles Lindbergh, hired by Henry Ford as a test pilot, used an adjoining airfield.[18]

Initially, the Ford Motor Company struggled to transfer automotive assembly practices to aircraft production at Willow Run. The use of steel-cast dyes hindered design changes to the bomber, and it was difficult to attract workers away from Detroit auto factories due to the distance and lack of local housing. Despite these issues, Willow Run was able to achieve remarkable production rates. At its peak in 1944, it produced a B-24 every

---

17 Hendricks, Jake. "Detroit Arsenal: Birthplace of American Tank Warfare." Military History of the Upper Great Lakes, October 11, 2015.https://ss.sites.mtu.edu/mhugl/2015/10/11/detroit-arsenal-birthplace-of-american-tank-warfare/.
18 "Encyclopedia of Detroit – Willow Run." Detroit Historical Society - Where the past is present. Accessed July 31, 2024. https://detroithistorical.org/learn/encyclopedia-of-detroit/willow-run#:~:text=Willow%20Run%20is%20an%20Albert,%24200%20million%20to%20the%20project.

hour. By 1945, it was able to produce 70 percent of its B-24s in two nine-hour shifts, with pilots and crew members sleeping on 1,300 cots as they waited for the B-24s to roll off the assembly line. At Willow Run the Ford Motor Company produced half of the B-24s out of the 18,493 that the United States built during the war.[19]

Billions of Bullets and Bombs

American industry was making impressive gains in building the weapons of war. A concurrent issue was making the mass of munitions required for them. Despite the overwhelming presence of the Dupont company in the domestic industrial landscape, the United States entered the World War II in poor shape concerning munitions. The stocks of ammunition on hand in 1940 were so meager that, in the words of Secretary of War Stimson, "We didn't have enough powder in the whole United States to last the men we now have overseas for anything like a day's fighting." To meet this situation, the U.S. Army Ordnance Department took steps in the summer of 1940 to create something new in American economic life—a vast interlocking network of ammunition plants owned by the government and operated by private industry. A great example of this is the Evansville Arsenal in Indiana. Originally a Chrysler automobile factory, it converted to produce .45 caliber cartridges (used in the M1911 pistol, M1 Thompson submachine gun, and the M3 "Grease Gun" submachine gun). From June 1942 to April 1944, the Evansville arsenal produced 96 percent of the military's .45 caliber cartridges: 3,264,281,914 rounds. The rejection rate of cartridges was less than .1 percent of production. Over 60 of these government-owned, contractor-operated plants were built between June 1940 and December 1942. Representing a capital investment of about $3 billion, they produced a wide range of military chemicals, loading millions of shells, bombs, grenades, rockets, and mines. The plants employed nearly a quarter of a million workers and covered a total land area equaling that of New York, Chicago, and Philadelphia combined.[20]

Surprisingly given the political tensions between the FDR administration and American industry before the war, the Nation showed a

---

19 Ibid.
20 Zimmerman, Dwight. "Bullets by the Billions: Chrysler Switches World War II Production from Cars to Cartridges." Defense Media Network. Accessed July 31, 2024. https://www.defensemedianetwork.com/stories/bullets-by-the-billions-chrysler-switches-world-war-ii-production-from-cars-to-cartridges/.

remarkable willingness to sacrifice and put all their efforts towards victory during the 1940s. A great example of this is the contributions of several key industry leaders who personally embraced the mission, sacrificing personal and institutional profits for the greater good. One of the most prominent is William Knudsen (1879-1948), who came to the U.S. as a poor immigrant from Denmark. He worked his way up the economic ladder by starting as a dockworker and eventually became an assistant to Henry Ford, the automobile manufacturer. Knudsen revolutionized mass production by building more flexible and efficient production plants during his work at Ford, Chevrolet, and General Motors, the last of which he was the CEO from 1937 to 1940. In 1940, President Roosevelt selected Knudsen to lead the nation's National Defense Advisory Council to prepare for war, and he became the first "Dollar-a-Year-Man" by leaving his $300,000-a-year CEO job to volunteer to direct the government industrial production effort. Knudsen made the Arsenal of Democracy possible as U.S. Director of Production by simplifying government contracting and repayment procedures, making it easier for corporations to produce for the war effort.[21]

The Sacrifice of the Nation

The demands of fighting a global war against two imperial powers put a heavy burden on American supplies of basic materials like food, shoes, metal, paper, and rubber. The Army and Navy were growing, as was the nation's effort to aid its allies overseas. Civilians still needed these materials for consumer goods as well. To meet this surging demand, the federal government took steps to conserve crucial supplies, including establishing a rationing system that impacted virtually every family in the United States. One of the WPB responsibilities was overseeing this effort, including planning and organizing events such as scrap metal drives, which were carried out locally to great success. For example, a national scrap metal drive in October 1942 resulted in an average of almost 82 pounds (37 kg) of scrap per American.[22]

Rationing was not limited just to materials, but also to what industry produced. From 1942 to 1946, no new Automobiles were made. For example,

---

21 "Manufacturing Victory: Who's Who in WWII Production" Accessed July 31, 2024. https://manufacturing-victory.org/history/ManufacturingVictory-factsheet.pdf.
22 "Material Drives on the World War II Home Front (U.S. National Park Service)." National Parks Service. Accessed August 5, 2024.https://www.nps.gov/articles/000/material-drives-on-the-world-war-ii-home-front.htm.

Ford built 691,455 automobiles in 1941, yet the company only built roughly 160,000 vehicles for civilians in 1942 before Ford's non-military car and truck lines ceased operations on February 10, 1942. The government then stockpiled remaining unsold cars and rationed them to those individuals deemed critical to public safety and the war effort—doctors, police and firefighters, farmers, and a rare handful of vital war workers. To be eligible for a new car, they had to possess an older car with more than 40,000 miles.[23]

The Fruits of American Labors

By the time the final Japanese surrendered in 1945, the United States had fulfilled President Roosevelt's admonition to become the great Arsenal of Democracy. American industry had produced more than 96,000 bombers, 86,000 tanks, 2.4 million trucks, 6.5 million rifles, and billions of dollars' worth of supplies to equip a truly global fighting force while maintaining a robust Home Front. The effects of this colossal effort far outlasted the war itself. Rather than returning to its depressed prewar state, the national economy added some twenty million new jobs over the next quarter century, doubling the size of the middle class.[24] More importantly, the infrastructure of factories, shipyards, supply chains, and trained personnel to produce those defense-related items was fully formed. After the war, many of these industries would shift back to purely commercial production, but by no means all of them.

Throughout much of American history, a post-war desire has been to dismantle war infrastructure and return to peacetime conditions in business and industry. This is perhaps best captured by the sentiments of President Dwight "Ike" D. Eisenhower at the end of his time in office. Ike served as the United States President from 1952 to 1960. Upon entering office, the retired Army General inherited the Korean War and an intense international competition with the Soviet Union. Despite the desire of both the previous Truman administration and Ike's administration to return defense spending to pre-World War II levels, global responsibilities compelled the United States to both reinstate conscription in 1948 and

---

23 Graff, Cory. "Making Automobiles Last during World War II: The National WWII Museum: New Orleans." The National WWII Museum | New Orleans, January 5, 2022.
24 "Becoming the Arsenal of Democracy: The National WWII Museum: New Orleans." The National WWII Museum | New Orleans, July 11, 2018. https://www.nationalww2museum.org/war/articles/becoming-arsenal-democracy.

keep defense spending levels high to field and maintain the military. In some of his final remarks as President, Ike cautioned against the pervasive and perhaps malign influence of the Military Industrial Complex in American democracy:

> *Until the latest of our world conflicts, the United States had no armaments industry. American makers of plowshares could, with time and as required, make swords as well. But we can no longer risk emergency improvisation of national defense. We have been compelled to create a permanent armaments industry of vast proportions. Added to this, three and a half million men and women are directly engaged in the defense establishment. We annually spend on military security alone more than the net income of all United States corporations…Now this conjunction of an immense military establishment and a large arms industry is new in the American experience. The total influence—economic, political, even spiritual—is felt in every city, every Statehouse, every office of the Federal government. We recognize the imperative need for this development. Yet, we must not fail to comprehend its grave implications. Our toil, resources, and livelihood are all involved. So is the very structure of our society.*[25]

Although the term "Military Industrial Complex" would acquire a slightly sinister reputation, Ike's warning did nothing to derail the size and strength of America's defense industries. Defense spending would hover between seven and nine percent of the total American Gross Domestic Product until near the end of the Vietnam War, and even then, stayed at or near five percent until the mid-1990s.[26]

However, the collapse of the Soviet Union and American military victory during Operation Desert Storm started to change American defense industries. Beginning in the President William J. Clinton administration, the United States government assessed that its military requirements had diminished and began actively reducing defense spending. Because many firms in the commercial Defense Industrial Base (DIB) heavily relied on the defense market, the Department of Defense determined that the sector's continued viability depended on restructuring. Accordingly, the U.S. government actively encouraged companies to pursue consolidation,

---

25 "President Dwight D. Eisenhower's Farewell Address (1961)." National Archives and Records Administration. Accessed July 31, 2024. https://www.archives.gov/milestone-documents/president-dwight-d-eisenhowers-farewell-address.
26 "U.S. Military Spending/Defense Budget 1960-2024." MacroTrends. Accessed July 31, 2024. https://www.macrotrends.net/global-metrics/countries/USA/united-states/military-spending-defense-budget.

with the result that by the early 2000s the number of prime contractors had diminished from fifty-one to five. Throughout the 1990s, the production output of the commercial DIB decreased by approximately 35 percent. The changing defense environment also affected government-owned capabilities, as the Department of Defense declared numerous facilities excess and closed them while reducing employment and activity at the remaining sites.[27]

Can We Ever Have the Arsenal of Democracy Again?

Today, the United States government defines the DIB as "the worldwide industrial complex that enables research and development, as well as design, production, delivery, and maintenance of military weapons systems, subsystems, and components or parts, to meet U.S. military requirements."[28] With the start of the Russo-Ukraine war in February 2022 and the massive requirements in munitions and combat systems generated by this conflict, many across the defense enterprise and industry have questioned whether the present American DIB is postured for success.[29] Many of the questions asked today are similar, if not the same, as those asked in 1917 and 1945, such as: How does American industrial Base gird itself for war? What is the role of Government versus the role of private industry? What is a realistic timeline to build industrial capability? Are there adequate sources of materials along with a trained and experienced workforce? What is the potential vulnerability and resilience of the DIB, especially in the case of a global conflict where adversaries can target the American homeland?

Historical comparisons are often, if not always, problematic. While it is understandable to harken back to the conviction that "we did it before and can do it again," it is critically important to understand how a past task was accomplished in the first place. Key legislation passed by Congress (or repealed in the case of the Neutrality Acts) in the years before

---

27 "The U.S. Defense Industrial Base: Background and Issues for Congress - October 12, 2023." US Congressional Research Service. Accessed July 31, 2024. https://crsreports.congress.gov/product/pdf/R/R47751.
28 "Defense Industrial Base Sector." Defense Industrial Base Sector | Cybersecurity and Infrastructure Security Agency CISA. Accessed July 31, 2024. https://www.cisa.gov/topics/critical-infrastructure-security-and-resilience/critical-infrastructure-sectors/defense-industrial-base-sector.
29 "The U.S. Defense Industrial Base: Background and Issues for Congress - October 12, 2023." Ibid.

direct American involvement in the war was foundational to building the Arsenal of Democracy. The government administrative structures like the War Production Board that existed only during the two world wars were essential to marshaling and controlling America's total resources. 80 years ago, the United States could only produce large numbers of tanks, trucks, ships, and planes by strictly restricting critical materials and significantly reducing the production of consumer goods. There was also a general and patriotic willingness to sacrifice on behalf of the war effort, matched by the inherent capability of American industry to respond to the demands of the Roosevelt administration. Calling the entire wartime military production effort "socialist" was not, for some, overstated.

Almost a quarter into the 21st Century, the United States Defense Industrial Base is amid a much-needed wake-up call. Spurred on by the ongoing Russo-Ukraine War and potential conflicts in the Pacific Rim, the Department of Defense and industry have begun to take concrete steps towards reinvigorating the DIB. In January 2024, the DoD released its first-ever Defense Industry Strategy, designed to guide specific actions and resource prioritization across the department.[30] Working in concert with its prime supplier, the Army hopes to double current 155mm artillery shell production by October 2024, with additional increases in capability after a new facility in Texas comes online.[31] These initial steps are far from the ultra-aggressive steps taken in the 1940s, but they do signal an acute awareness of the challenges ahead. Is there a once-and-future Arsenal of Democracy for America? Does the nation have both the political will and industrial capability to fight and win a conflict that could consume munitions, supplies, systems, and people at a rate not seen since 1945? The answer to these questions could define the world's future history for the next 80 years and beyond.

---

30 "DOD Releases First Defense Industrial Strategy." U.S. Department of Defense. Accessed July 31, 2024. https://www.defense.gov/News/News-Stories/Article/Article/3644527/dod-releases-first-defense-industrial-strategy/.
31 Skove, Sam. "Army Aims to Double 155mm Shell Production by October." Defense One, February 6, 2024. https://www.defenseone.com/policy/2024/02/army-aims-double-155mm-shell-production-october/393943/.

4

STRATEGY SHORT OF WAR

Ron Granieri

Logistics is about answering five fundamental questions: Who, What, When, Where, and How? Successful logistical management is a matter of figuring out who needs to deliver what to whom, at what place, at what time, for what purpose, and by what method. Most of the time when military analysts discuss logistics, they think of their relationship to the battlefield, about bringing supplies up close enough to the front so that the warfighter can use those supplies to do the voodoo that warfighters do. But logistics can also be decisive when there are no rockets' red glare or bombs bursting in air. Indeed, when managed appropriately, logistics can provide opportunities for strategic success without resorting to open hostilities. One especially instructive example of the geopolitical power of logistics below the level of armed conflict is the Berlin Blockade and Airlift from June 24, 1948, to May 12, 1949.

The Berlin blockade was one of the greatest logistical accomplishments in history. For nearly a year, American and British pilots supplied a major European city with food, fuel, and virtually all the other necessities of life through the air. According to an official State Department history, "at the height of the campaign, one plane landed every 45 seconds at Tempelhof Airport."[1] The logistical accomplishment is even more impressive when one considers that this all happened before the advent of jet aircraft or giant transport planes. The Airlift required the meticulous organization of World War II-era aircraft, mostly C-47s and C-54s, taking off from bases in western Germany, landing in Berlin, and returning almost immediately. The successful airlift shaped the future not only of Berlin and Germany, but of the Atlantic alliance and the Cold War itself.

With all that in mind, any conversation about logistics must include the story of the Berlin airlift, which is full of historical significance and irony.

---

1   https://history.state.gov/milestones/1945-1952/berlin-airlift.

The Unique Situation of Postwar Berlin

Students new to Cold War history often incorrectly conflate the Berlin blockade with the building of the Berlin Wall. Those are two separate events separated by more than a decade, though both reflect the anomalous historical and geopolitical reality of Cold War Berlin. After defeating Nazi Germany in 1945, the victorious Allies (the United States, Great Britain, France, and the Soviet Union) formally dissolved the German government and assumed joint responsibility for managing German affairs. They divided the defeated nation into four occupation zones, within which each power would have wide-ranging autonomy over political, economic, and administrative development, though policy was notionally to be coordinated and harmonized by an Allied Control Council (ACC) based in Berlin. The former *Reich* capital was, however, in the middle of the Soviet occupation zone. To address this issue, the Allies divided the city into four occupation zones. Thus, each occupying power was at home in Berlin.

This arrangement (which was mirrored on a smaller scale in the Austrian capital, Vienna) reflected wartime planning that was itself based on the assumption of continued Allied cooperation in applying the "Four Ds" (Denazification, Demilitarization, Decentralization, and Democratization) to Germany. Such cooperation found expression in early efforts such as the 1945-46 trials of the major war criminals in Nuremberg but relatively quickly foundered on the very real differences of vision among the occupying powers for the future of Germany and Europe. Put most simply, there was significant disagreement over whether the goal of the occupation was primarily punishment or rehabilitation of Germany. The Soviets (and to a lesser extent, the French) viewed the occupation as an opportunity to extract reparations to pay for the destruction wrought by Nazi imperialism. The Americans and British, however, drawing on their understanding of the failures of interwar policy, wanted to integrate Germany into the broader European and world economy to help with recovery and avoid a repeat of the catastrophes of the 1930s.[2]

The contrast between these approaches grew during the first year of the occupation. As the Soviets literally dismantled much of the industrial base of their zone to use for the reconstruction of their shattered economy,

---

2 For broad discussion of the division of Germany, see John Lewis Gaddis, We Now Know: Rethinking Cold War History (Oxford UP, 1997), especially 113-135 and, more critical of the Anglo-Americans, Carolyn Woods Eisenberg, Drawing the Line: The American Decision to Divide Germany, 1944-1949 (Cambridge UP, 1996).

the Americans and British declined to impose reparations, arguing that each occupying power could only act within their zone. A series of meetings of the allied foreign ministers could only paper over the differences, and that not for long. Divergent policy priorities became especially apparent by 1947. At the start of the year, the British and Americans formally combined their occupation zones into a single administrative unit (dubbed "Bizonia") to encourage reconstruction of a market economy in Germany. In March, President Truman announced the Truman Doctrine, which promised American support for free peoples struggling against totalitarianism. Then in June, Secretary of State Marshall announced the European Recovery Program (the Marshall Plan) of American assistance for European recovery. Marshall declared that American aid was open to all, but it depended on Europeans developing economic cooperation plans that would break down bottlenecks and encourage overall development. The Soviets saw all these developments as proof of American desire to impose their system on Europe, and after failing to gain a promise of reparations payments with no strings attached, not only refused to participate in the Marshall Plan but also forbade those eastern European states within the Soviet sphere from participating either.

The consequences for Germany were not far to seek. As the British and Americans (along with, eventually, the French) included their German occupation zones in the Marshall Plan, the Soviets kept their zone out, spurring momentum for the *de facto* if not *de jure* division of the country.

The Blockade Begins

Events came to a head in 1948 with the emergence of Western plans to reform the German currency. The *Reichsmark* had become virtually worthless by the end of the war. In much of Germany, economic life revolved around black-market trading or barter with American cigarettes or other luxury goods. If Western plans to link German recovery to overall European recovery were to have any hope of success, Germany needed a new currency that people would trust. Accordingly, the U.S. Mint began to produce stacks of new notes and coins for Operation Bird Dog, the code name given for the switch of currencies. As plans for introducing the new *Deutschmark* progressed through the Spring, they ran up against Soviet refusal to participate. The Soviets were, unsurprisingly, not interested in having their occupation zone drawn into a Western economic model. Furthermore, considering the porous borders between occupation zones in Berlin, they also opposed introducing the new currency in the jointly occupied capital city.

Hoping to shift the terms of the debate to their political advantage, the Soviets initially framed their arguments against the currency reform in the language of German unity. The Allies had promised to govern Germany collectively, so it made no sense for Germany to be divided economically. Although the Soviets could not offer any specific alternative that would be acceptable to all sides, they argued for renewed negotiations and a delay in any irreversible steps, trying to portray themselves as the force for moderation.

Even as they claimed to reject confrontation, the Soviets sought to pressure the Allies by making life in Berlin more difficult. After walking out of the ACC in March, the Soviets began to harass and limit access to Berlin, playing hardball within existing legal limits. Berlin was deep in the Soviet zone, and the Soviets controlled the land and water routes to the city. They could slow traffic in the name of searching for contraband or escaped prisoners, or stop it altogether, without necessarily breaking any agreements. Temporary halts to traffic in April and May, along with public speculation that the Soviets might impose a new currency of their own on their Zone and in the city, sent a message that the Western position in Berlin was in jeopardy if the Allies continued with their plans.

Western leaders were painfully aware of their precarious position. As Robert Murphy, the Political Advisor to the American military governor, wrote to a State Department colleague, "The Berlin situation is peculiarly difficult, surrounded as Berlin is by the Soviet zone. If the SMA [Soviet Military Administration] decides to issue a new currency for the zone, it will be, according to the technicians, practically impossible for the Western sectors of Berlin to operate on an independent basis."[3]

The ball was now in the Allied court, as the Allies had to come to the negotiating table or do something else. American military governor Lucius Clay initially argued for calling Moscow's bluff, sending an armored column along the highway to Berlin, while Winston Churchill advocated everything from harassing Soviet shipping in the Suez and Panama canals

---

3 The United States Political Adviser for Germany (Murphy) to the Assistant Secretary of State for Occupied Areas (Saltzman), 13 April 1948. Foreign Relations of the United States 1948, Germany and Austria, Vol. II, Document 532 https://history.state.gov/historicaldocuments/frus1948v02/d532.

to threatening with atomic weapons.[4] Such suggestions did not find much support in Washington, where no one wanted to take such an extreme risk, though all agreed "that we can only arrest and deter [the Soviets] by a real show of resolution."[5] An awareness of that risk aversion also worried Murphy, who said in a later note, "The only feature, frankly, which worries me about our Berlin position is strength of determination in Washington to maintain the position. What I fear is that, in view [of the] unfavorable logistics [of] this situation, its expenses, and…lack of specific purpose for our presence here may over the longer term combine to vitiate US determination to hold fast."[6] Nor were the Allies completely in agreement on what to do. French representatives were particularly unenthusiastic about holding onto Berlin and "would not go into mourning…if Berlin were abandoned."[7]

When the Soviets made their *fait accompli* complete with the imposition of the blockade on June 24, the Allies had to decide on a course of action. If they were to maintain the Western position in Berlin but did not want to provoke an escalation, they had to find a way within existing agreements to sustain the western zones. Since there were written agreements guaranteeing air corridors to Berlin, an airlift was an option, though no one knew if it would be possible to sustain the city for long through the air. On June 26, General Clay and General Curtis LeMay, the commander of U.S. Air Forces in Europe, announced initial plans for airlifts between the Western Zones and Berlin based on an estimated daily supply of 225 tons, with about 70 planes at their immediate disposal. This tonnage would be barely enough for the immediate needs of the occupation forces, though the needs of the civilian population were set at nearly 2,000 tons daily.[8] Recognizing the open-ended challenge, Secretary of State Marshall nevertheless declared that the allies would "supply the city by air as a beleaguered garrison."[9]

---

4   The Ambassador in the United Kingdom (Douglas) to the Under Secretary of State (Lovett), 17 April 1948. Ibid, Doc 536. https://history.state.gov/historicaldocuments/frus1948v02/d536.
5   Ibid.
6   The United States Political Adviser for Germany (Murphy) to the Secretary of State, 13 April 1948. Ibid, Doc 533. https://history.state.gov/historicaldocuments/frus1948v02/d533.
7   The United States Political Adviser for Germany (Murphy) to the Secretary of State, 15 April 1948. Ibid, Doc 534. https://history.state.gov/historicaldocuments/frus1948v02/d534 See also The Ambassador in France (Caffery) to the Secretary of State, 24 June 1948. Ibid, Doc 556. https://history.state.gov/historicaldocuments/frus1948v02/d556.
8   The United States Political Adviser for Germany (Murphy) to the Secretary of State, 26 June 1948. Ibid., Doc 558. https://history.state.gov/historicaldocuments/frus1948v02/d558.
9   The Secretary of State to the Embassy in the United Kingdom, 28 June 1948. Ibid. Doc 564. https://history.state.gov/historicaldocuments/frus1948v02/d564.

The Allies had decided on the what and the where; the question of how and for how long remained to be answered.

The Airlift in Practice

No one had ever devised a workable plan to provision a major city purely from the air, so this was a huge logistical challenge. Also, once the allies accepted the challenge, they committed their prestige to the effort. Once begun, "Operation Vittles" had to be seen through to the end, and no one quite knew where that end would be. They also had to hope that the Soviets would not choose to escalate by interfering with the air traffic to Berlin. On the first day, 32 American C-47s delivered 80 tons of provisions while flying to the only two airports in the western zones, Tempelhof in the American Zone and Gatow in the British. Major General William Turner commanded the operation, declaring that he wanted the airlift to operate in "rhythm, on a beat as constant as a jungle drum."[10]

It took time for the rhythm to develop, but the growth was continuous. By August 1949, the combined British and American flights had crossed the 4,500 tons daily threshold considered necessary for the city to survive. By September 1948, the airlift brought 7,000 tons of supplies daily; by February 1949, more than 8,000 tons were arriving daily on over 900 flights; the month of March 1949 saw 200,000 total tons delivered. After the Soviets lifted their blockade on May 12, 1949, flights continued to stock up Berlin's reserves. The 276,926th and final flight of the airlift arrived in Berlin on September 30, 1949. All told, nearly 700 aircraft logged over 124 million miles and delivered roughly 2.3 million tons of supplies.[11]

The logistical accomplishments also included major infrastructure projects. A new concrete runway at Gatow was built in a matter of weeks. Even more impressive, the Allies constructed a brand-new third airport in the western sectors, Tegel in the French zone, in about 90 days. Tegel then went on to a seven-decade career as Berlin's main passenger airport until it was replaced by the new Berlin-Brandenburg airport, which required more than a decade to complete.

Of course, reality is more complex than triumphalist narratives sometimes allow. Berlin's black market continued to hum and smuggling

---

10 Turner quote and other data are drawn from "The Berlin Airlift Timeline," The American Experience. https://www.pbs.org/wgbh/americanexperience/features/airlift-berlin/.
11 Ibid. For more details, see also Barry Turner, The Berlin Airlift: The Relief Operation that Defined the Cold War (Icon Books, 2017).

between the Soviet zone and West Berlin continued.[12] And there was a significant human cost. An estimated 65 people died over the course of the airlift due to crashes and accidents.

Special human grace notes also marked the airlift, including both Operation Santa Claus, which delivered Christmas gifts to thousands of Berlin's children, and especially the exploits of the "Chocolate Pilot" Gail Halvorsen and his colleagues who dropped small parachutes with candy to eager children.[13] Stories of Berliners expressing their gratitude for the planes delivering their payloads are even more fascinating when one considers that many of those same Berliners had dreaded the sound of allied aircraft only a few years before—and that those aircraft had been guided by many of the same pilots, delivering much less friendly payloads.

The transformation of Berlin from the capital of a hated adversary to a front-line city of brave and embattled men and women is one of the most important elements of the Airlift. In September 1948, West Berlin's Mayor, Ernst Reuter, addressed a cheering crowd of more than 300,000 Berliners and called on the world to "Look at this City!" (Schaut auf diese Stadt!), drawing attention to their heroism and sacrifice. Reuter became a hero not only to his constituents but to people the world over, profiled in magazines such as *Life* as part of a group of leaders who were going to bring Germany into a new era of freedom and democracy.[14] If the Soviets had hoped to stop the creation of a free and democratic Germany, or to break the ties between the Western Allies and Berlin, their plans had backfired spectacularly.

Conclusion—The Legacy of the Airlift

What can we learn from the history of the Berlin Airlift? As we said at the beginning of this essay, logistics are about who, what, where, when, why, and how. Successful logistical management delivers what you need, where

---

12 Paul Steege, Black Market, Cold War: Everyday Life in Berlin, 1946-1949 (Cambridge UP, 2007).
13 Andrei Cherny, The Candy Bombers: The Untold Story of the Berlin Airlift and America's Finest Hour (Dutton, 2009).
14 "Ihr Völker der Welt, schaut auf diese Stadt!" – Ernst Reuters Rede vor dem Reichstag https://www.swr.de/swrkultur/wissen/archivradio/voelker-der-welt-schaut-auf-diese-stadt-ernst-reuters-rede-vor-dem-reichstag-100.html. See also Brian C. Etheridge, From Enemies to Allies: Cold War Germany and American Memory (University Press of Kentucky, 2016), 85-93.

you need it, when you need it, and to use how you need it. The case of the Berlin Airlift shows how getting logistics right has a significance that can be worth many divisions.

The Allied ability to organize the airlift and follow through on their political promises not only saved the Western position in Berlin, but also demonstrated resolve in a time of crisis and uncertainty. Robert Murphy and his colleagues had been acutely aware of the tenuousness of their position in Berlin. Being able to assert that position without resorting to armed conflict scored a major propaganda victory and sent a clear message to the Soviets that served an important political purpose in the early years of the Cold War.

That resolve also had a profound impact on the German public and German-American relations. Residents of Berlin did not forget the generosity expressed by the United States. They understood the sacrifices being made and the symbolism of what America was doing to help a city on the front lines of the long twilight struggle. Germans and Americans both looked at each other with new eyes because of the Airlift. The consolidation of the western occupation zones into the Federal Republic of Germany by the summer of 1949 was advanced by the common exertions of the airlift and began a new era for Germans and their relationship with their neighbors.

Operation Vittles helped save a city and was one of the greatest logistical triumphs of the 20th century. More than that, it was a strategic victory for the West that was accomplished without firing a shot. That's what effective logistics can do.

5

THE FOUR LOGISTICAL OPERATIONS

Richard Killblane

Logistics is boring—until you run out, but it defines the conduct of war. Strategy defines how to win wars, and operational art is the study and conduct of campaigns. Campaigns represent a series of movements and battles to achieve an objective. However, to truly comprehend operational art, one must understand logistics and its effect on military operations. In this light, the three different functions of logistics (transportation, supply, and service) define the conduct of campaigns. While the maneuver units can travel over almost any kind of terrain, they must be supplied. Therefore, lines of communication (supply lines) dictate the avenues of approach or the maneuver box in which units fight. Days of supply control the rates of advance no matter how well trained or aggressive the maneuver force is. When the Third Army ran out of fuel racing across Northern France in 1944, it had to halt. Likewise, when the 3rd Infantry Division ran out of bottled water in its race across the desert to Baghdad in 2003, it came to two operational pauses until it built up its appropriate days of supply of water to advance again. The underappreciated service function keeps soldiers and their equipment in the fight. Services include maintenance, cooking, laundry, field showers, and medical services. These functions combine into logistical operations.

Similar to maneuver, movement creates logistics operations, which coincide with a myriad of military operations, whether combat or other than war. The first, strategic deployment, delivers the military force to the theater of operations. Reception, staging, onward movement, and integration (RSO&I) move the arriving units through the funnel of air and seaports of debarkation to their staging areas. Sustainment begins with the first boots on the ground, and as described, commanders develop their operations based on a balancing act of achieving the military objective with the ability to sustain the force in theater. Retrograde is RSO&I and strategic deployment in reverse but compounded by the progressive accumulation of people, supplies, equipment, battle damage, and political constraints. It does not matter what the military operation is, whether the offensive drive to Baghdad or earthquake relief in Haiti, the four logistics operations always shape the military operation and operational art. This study will explore how logistics operations have evolved and affected military operations since the Spanish-American War.

Strategic Deployment

When military historians analyze why one side won a battle against another, they usually compare which side committed the most combat power and how they employed it. With the exception of instances of military genius, the side with the most soldiers usually won. Unfortunately, students of war rarely ask how those soldiers arrived in the theater of operation. Unless invading a neighboring country, deploying a long distance around the globe is a very difficult task. Therefore, deploying from home station to a theater of operation is critical to achieving mass on the battlefield, and to fully comprehend the full scope of war requires a study of strategic deployment. The objective of strategic deployment is to deliver as much combat power, support, and sustainment as possible to win battles and, ultimately, the war. Because the United States is separated from its potential adversaries by two vast oceans, it has invested much time and effort perfecting how to deploy superior combat power around the globe.

In the United States, the planning process begins with the president, who defines the end state, which prior to 1947 the War and Navy Departments and later Department of Defense and combatant commands translated into strategic objectives. The theater commander then develops the strategy to achieve military objectives, and the number of combat forces available in theater has been a key factor in developing the strategy. Furthermore, how many combat units one can deploy into a theater in the shortest time shapes the strategy and subsequent campaigns, and strategic deployment is consequently measured by speed and volume.

Commencing with the Spanish-American War, all American wars would be fought overseas, inherently requiring strategic transportation. This new era of strategic deployment started with the assembly and deployment to Cuba, which included severe growing pains. The strategy called for deploying 25,000 soldiers of Major General William Shafter's

Fifth Corps from Tampa, Florida, to Santiago, Cuba, before the Yellow Fever season began in July 1898. Regular Army regiments immediately began traveling to Tampa by rail and would continue to arrive as late as June 2. To transport these soldiers to Cuba, the Quartermaster Department had purchased or leased 31 vessels by May 26. The calculation for bunk space per ton of cargo space turned out to be wrong, and after the construction of the bunks, the fleet could only bunk 17,000 soldiers, a serious shortcoming. In addition, the Port of Tampa usually supported day trips to Key West, not uploading mass quantities of cargo. The stevedores had no material handling equipment and had to load the cargo by hand. This issue delayed the upload until June 6. The following day, Shafter, who had no experience with this type of massive movement, warned that any deploying regiment would be left behind if it was not at the port the following day. To no surprise, this warning created a race where regiments commandeered vessels to ensure their transport. Somehow, the Quartermaster at the port, COL Charles Humphrey, managed to find space for everyone, but an unconfirmed sighting of the Spanish fleet delayed the American invasion fleet's departure until June 14, 12 days behind schedule. The Fifth Corps would not charge up San Juan Hill until July 1, and then it had to lay siege for 17 days. While the corps won the campaign in Cuba, the feared diseases, such as Yellow Fever, caused many deaths, some of which may have been prevented by a faster strategic deployment.[1]

Anything that could have gone wrong had. The Army also had to deploy troops to the Philippines and could not afford another failure like the one at Tampa. What the Army needed was professionals who understood strategic deployment. In 1899, the Quartermaster created the Army Transportation Service (ATS) with the authority to coordinate directly with railroad officials and port authorities, and the ATS also managed the Army's fleet of transport vessels. With practice, the Army would master strategic deployment.

By the time of American entry into World War I in April 1917, the War Department created the Embarkation Service under the new General Staff to coordinate the movement of troops, equipment, and supplies overseas. This organization coordinated transportation to the ports of embarkation at New York City, New York; Hoboken, New Jersey; and Newport News, Virginia. It would schedule men and their equipment to arrive by rail when

---

1 Richard E. Killblane, *Delivering Victory; The History of U.S. Military Transportation*. (Bingley: Emerald Publishing, 2020), 12-14.

their transports became available. The Embarkation Service inherited a monumental task, as the War Department initially planned to create five divisions out of Regular Army regiments and two of the U.S. Marine Corps, 16 divisions from the National Guard, and 17 National Army divisions from conscripts. The General Staff wanted to deploy the Regular Army divisions first, then the National Guard since they already had limited training during service on the Mexican border, and finally the National Army divisions since draftees required the most time for training. Eventually, the General Staff concluded that it would require 80 divisions to defeat the German Army.[2]

While the remaining divisions trained for war, major traffic jams impeded commercial transportation. Rail represented the first leg of deployment to the seaport of embarkation. However, the lack of centralized control imposed upon rail by the Interstate Commerce Act of 1887 led to a lack of cooperation and logistics jams accumulating outside the ports. Consequently, Congress created the United States Railway Administration with the Railway Administration Act of March 21, 1918. The Railway Administration oversaw the production of new rolling stock, standardized locomotives, and set prices to keep costs down. The American Railway Association also created a Troop Movement Section to move units from their training bases to the ports. To compound the problem of traffic jams, severe winter storms from December 1917 through February 1918 froze the New York harbor, impeding the ability of ships to sail. However, the major limitation of strategic deployment was insufficient ocean-going transport.[3]

At the outbreak of war, the ATS owned just enough ocean transports to support deployments and sustainment of a small Regular Army in Hawaii and the Philippines. In September 1916, Congress authorized the creation of the United States Shipping Board to exercise control over shipping and the construction of vessels to enable the merchant marine fleet to meet the demands of the war in Europe. Unfortunately, the shipyards could not complete the construction of the transports needed to ship troops, equipment, and war supplies to France before the war ended. Not only that, but there were not enough ships available to buy or lease to transport the massive number of divisions believed necessary to win the war. Fortunately, 21 German vessels were anchored in neutral U.S. ports, and the confiscation of these added significantly to the Army's transport

---

2 Killblane, *Delivering Victory*, 17 and 27.
3 Killblane, *Delivering Victory*, 17-18.

fleet; however, their confiscation created a manning problem. By January 1918, the U.S. Navy Bureau of Operations created the Naval Overseas Transportation Service (NOTS) to recruit and train crews for the Army vessels. Since the Navy owned warships, it retained the responsibility to escort the convoys overseas. Assembling convoys also took precious time. All these measures still proved insufficient to both deploy its army to France and supply its allies.[4]

This requirement to supply its allies with a limited number of transports significantly hindered the deployment of combat divisions and appropriate supporting units. The British offered to use its sea-going fleet to deliver American divisions to augment the Allied armies. Fortunately, the fresh American divisions gave the Allies an advantage over their war-weary opponents, and the Allies launched a major offensive in October that led to an end of hostilities on November 11, 1918. This proved fortuitous as the lack of ocean transports could not deliver enough support units to support the planned 80 American combat divisions needed to win the war in France. It had only delivered 43 divisions by the time of the Armistice. The Achilles heel of strategic deployment in the two previous wars remained sealift. The rush to go to war did not allow time to construct adequate transports. The next war would prove different.[5]

During World War II, the United States created the largest military force in its history, which fought in several theaters of operations around the globe. The newly created Office of the Chief of Transportation became responsible for planning, coordinating, and moving men and materiel from their origin to ports of embarkation and then overseas. This organization built on the successful lessons of the previous war. The Army had priority for rail movement, so upon completion of training and equipping, units moved by rail to a staging camp adjacent to one of the ports of embarkation (POE). New York, New York, and Hampton, Virginia, POEs deployed units to Europe; New Orleans, Louisiana, POE deployed units to the Caribbean; and San Francisco, California, POE deployed units to the Pacific and India. Just as had happened during the previous war, the number of ocean-going transports on hand limited the number of units that could deploy. So, the priority went to winning first in North African and Mediterranean Theater and then European Theater while the Pacific Theater would fight a strategic defense. After victory in Europe, forces would then shift to the Pacific

---

4 Killblane, *Delivering Victory*, 19-20.
5 Ibid., 20.

Theater. Part of that defensive strategy also involved supplying Chang Kai Chek's Chinese National Army to contain the Japanese in the China Burma India Theater.[6]

Before the United States declared war, it barely had enough transports to deliver sufficient forces to win in any theater. In September 1941, the Army Transport Service could transport 18,000 troops and 177,000 tons of cargo, respectively. The Naval Transportation Service also had a lift capacity of 35,000 troops and 273,000 tons of cargo. At the request of President Franklin D. Roosevelt, the U.S. Maritime Commission modified the design of its standardized C cargo ships to expedite construction. After 244 days of construction, the first of these new vessels, SS *Patrick Henry*, was commissioned in September 1941, and the new ship type became known as Liberty Ships. As many as 18 shipyards built these ships, and as experience improved, the average construction time reduced to 42 days. However, the new Liberty Ships did not immediately solve the shipping shortcomings. Also, German U-boats were sinking vessels at an alarming rate until advances in anti-submarine warfare made passage across the North Atlantic safer after October 1943.[7]

To counter the German U-boats' success, the newly formed U.S. War Shipping Administration began building Victory Ships, which were larger and faster than Liberty Ships and, more importantly, the German U-boats. Six different American shipyards built these Victory Ships, the first of which was completed in February 1944. By war's end, shipyards had constructed 2,710 Liberty Ships and 531 Victory Ships, an incredible feat. Compared to the 1,200 combat ships of the U.S. Navy, the Army earned the reputation for having more ships than the Navy.[8]

With the limited number of transports available in the beginning, the Office of Transportation loaded men and materiel bound for a staging base in England, Hawaii, or Australia. Once the units arrived, they would continue training. The amphibious invasion of North Africa, however, was launched directly out of Hampton POE in October 1942. There were two ways to load vessels: pack everything that would fit or combat load. The latter arranged the landing force and follow-on support in the order needed on shore. This effort required detailed planning, and the War Department

---

6   Killblane, *Delivering Victory*, 33-34.
7   Benjamin King, Richard C. Biggs, and Eric R. Criner, *Spearhead of Logistics; A History of the U.S. Army Transportation Corps*. (Fort Eustis: US Army Transportation Center, 1994), 138-139; and Killblane, *Delivering Victory*, 34-35.
8   Killblane, *Delivering Victory*, 34-35.

sequenced the arrival of supplies based on consumption tables for given units over a specified time and the best-guess approach that seemed inflexible. This pattern of detailed planning continued with the amphibious operations out of North Africa and England.

By the end of the war in August 1945, the Army had deployed 61 divisions to Europe organized under the First, Third, Seventh, and Ninth Armies, 15 divisions to the Mediterranean under the Fifth Army, and 22 divisions to the Pacific organized under the Sixth, Eighth and Tenth Armies. This effort was a phenomenal feat accomplished by the massive shipbuilding effort in the United States. The United States Armed Forces had never moved that many units overseas and never would again. To avoid rebuilding a transport fleet, the Army mothballed its fleet of Liberty and Victory Ships, known as the Reserve Ready Fleet, in Virginia, Texas, and California. With the reorganization of the Department of Defense in 1947, the Army turned over its fleet of transports and freighters to the Navy's Military Sealift Transport Service.

The war left the United States Army on occupation duty in Germany and Japan, and the Cold War soon followed. The United States allied with European and Pacific partners of the free world to deter any aggression by the Soviet Union and Communist China. Not five years after the end of the previous war did the United States enter another, this time in Korea when the communist North Korean People's Army invaded South Korea in June 1950. Fortunately, four of the six Army divisions that would initially deploy to that war came from nearby Japan. Only the 1st Marine Division, and the 2nd and 3rd Infantry Divisions deployed from the Continental United States, so sealift was not an issue. The Eighth Army and X Corps acquired enough combat power to beat back the North Koreans and then contain the Chinese along the prewar boundary. The real challenge occurred during the deployment for the next war.[9]

When General William Westmoreland escalated the ground war in the Republic of Vietnam in 1965, he asked President Lyndon B. Johnson for half a million men. Not only did the Army not have that many soldiers, but neither did it have enough sealift to deploy them because of competition for sustaining the Seventh Army in Germany and Eighth Army in Korea. The war planners in the U.S. Pacific Command calculated that it had

---
9 Richard E. Killblane, "Operation Yo-Yo: Transportation during the first year of the Korean War," US Amy, October 8, 2013, Operation Yo-Yo: Transportation during the first year of the Korean War | Article | The United States Army.

enough sealift to deploy increments of 125,000 troops over three years from 1965 to 1967. This incremental buildup therefore dictated the strategy. The first combat units to arrive defended the depots at the ports, except the 3rd Brigade of the 25th Infantry Division at Pleiku and the 1st Cavalry Division at An Khe. The second increment would expand military operations from the depots, and the third increment would focus on closing the gaps between them. This approach became known as the "enclave concept."[10]

Army strategic deployment to Vietnam began at the units' home installation. Military Traffic Management and Terminal Service (which became Military Traffic Management Command in 1973) was responsible for ground transportation and operation of the military ocean terminal. At the time, interstate highways made buses the primary mode of ground transportation for personnel, although rail remained the primary mode for equipment. Some personnel also traveled across the country by air. In the first two years, units met their equipment at Oakland Military Ocean Terminal in California, and the soldiers and equipment sailed together to Vietnam. Interestingly, this transoceanic transport was aboard the Liberty and Victory Ships the Soldiers' fathers and uncles had sailed on during World War II. By the third year of the Vietnam War, the Army changed the deployment process by sailing the equipment and flying the men by commercial air. From 1967 onward, this method was the norm. Overall, the Armed Forces deployed eight complete divisions (one of those a Marine Corps division), seven separate infantry brigades, an armored cavalry regiment, and an aviation brigade. These combat units fell under three corps-level headquarters during the first three years of the war. The deployment to Vietnam had been very slow, but afterwards the Armed Forces would revolutionize strategic deployment into an art.[11]

Since WWII, aircraft have become capable of transporting increasingly larger numbers of passengers, making it the preferred method for transporting personnel. The two key factors to deployment were speed and volume. Aircraft could deliver soldiers faster, but ocean-going ships could carry more materiel. This capability inspired the Rapid Deployment Force capability, where light infantry and airborne units could deploy within 18 hours of notification. This was no easy task, as it required regular practice in the form of emergency readiness deployment exercises (EDRE).

---

10 Killblane, *Delivering Victory*, 131-132 and 135.
11 Killblane, *Delivering Victory*, 128 and 132-133, and Shelby L. Stanton, *Vietnam Order of Battle*. (New York: Galahad Books, 1986).

The invasions of Grenada in October 1983 and Panama in December 1989 validated this concept. Two Ranger battalions and a brigade of the 82nd Airborne Division descended on the tiny island of Grenada on D-Day while a brigade of the 82nd Airborne Division, brigade, and division headquarters from the 7th Infantry Divisions (Light), and all three Ranger battalions and their regimental headquarters deployed from five different installations to arrive on D-Day. The United States could then project force with speed and surprise around the globe.

Using this concept, the 82nd Airborne Division was able to arrive first in Saudi Arabia during Operation Desert Shield in August 1990. For large-scale deployments by air that exceeded the U.S. Air Force's capability, the Department of Defense had an agreement with commercial carriers, known as the Commercial Reserve Air Fleet (CRAF), to deploy troops with commercial aircraft. CRAF aircraft served the Army well during routine deployments to the Sinai as well as to Saudi Arabia during Desert Shield and later Operations Enduring Freedom (OEF) and Iraqi Freedom (OIF). During OIF, the U.S. Transportation Command deployed by air the personnel of four divisions and supporting units to Kuwait and subsequent annual rotations of troops. Coincidently, the shutdown of commercial flights after the terrorist attack on September 11, 2001, made plenty of commercial aircraft available. Rapid deployment was required for emergency deployments, but if the combatant command anticipated a threat, then the deployment could be very slow and gradual.[12]

Infiltration provided an additional means of deployment to build mass. The escalating situation in Panama that started in March 1988 led to several buildups. The first show of force responded with the deployment of attack and lift helicopters out of the 7th Infantry Division (Light) in March with a continued rotation of personnel. The subsequent war plans required the deployment of two more brigades from the 7th Infantry Division and a mechanized infantry battalion from the 5th Infantry Division (Mechanized) to augment the 193rd Separate Infantry Brigade already stationed in Panama. While the Army designed the light infantry to deploy rapidly, it would take weeks to deploy the mechanized infantry battalion by fast sealift. As a result of the aborted elections in May 1989, U.S. Southern Command (SOUTHCOM) asked for the deployment of one of the light infantry brigades and the mechanized infantry battalion. All SOUTHCOM needed to complete the plan was one more light infantry brigade. When

---

12 Killblane, *Delivering Victory*, 252.

the XVIII Airborne Corps changed the plan in October 1989, it replaced the brigade from 7th ID. (L) to an airborne brigade out of the 82nd Airborne Division, it also added the need for M-551 light tanks. Army watercraft secretly infiltrated those tanks to Panama so they would be in place when the Panamanian Defense Force killed Marine 1st LT Robert Paz on December 16, 1989. Everything else needed to complete the deployment could arrive at H-hour. Since transportation had always been a limited factor, the next revolution came in how to plan and coordinate these strategic assets.

Facilitating rapid deployment planning required a standardized process. The deployment planning during the Vietnam War inspired the Joint Chiefs of Staff to standardize the planning process into the Joint Operating, Planning System (JOPS) in 1970, which evolved into the Joint Operations, Planning, and Execution System (JOPES) in 1979. This process culminated in the Time Phase Force Deployment Data (TPFDD), which produced a detailed timetable predicting each unit's departure and the exact airframe or vessel it deployed on to its arrival in theater. These systems ensured that the required modes of transportation were available to deploy specific units in the agreed upon sequence of deployment into the next century. Computerization would expedite the process even further.[13]

The Armed Forces began working with computers during the 1960s, but the computer revolution during the 1980s significantly improved deployment planning. The transition from vacuum tubes to transistors and microchips reduced the size of wall-to-wall UNIVAC computers of the 1960s to desktop and handheld computers during the 1980s. The modem allowed computers to talk to each other in messages of ones and zeros. Subsequently, different transportation organizations began developing programs to rapidly process and exchange information. Automation increased the speed and efficiency of planning exponentially. During the 1980s, the Department of Defense also modernized its electronic communications system into the Worldwide Military Command and Control System (WMCCS) (a military predecessor to the Internet), which allowed participants to disseminate information instantaneously around the globe. As the systems improved, deployment planning became a more efficient process of loading data into a computer program connected with all the participants in the deployment process.[14]

This technical revolution of the 1980s allowed deployment planners to develop a series of additional computer programs that tracked the flow

---

13 Killblane, *Delivering Victory*, 156 and 171.
14 Killblane, *Delivering Victory*, 169-171.

of passengers and cargo but required someone to manually load data into the program for others to know where the cargo was. In 1978, the Navy pioneered the Navigation System with Timing and Ranging (NAVSTAR), which became the global positioning system (GPS) that accurately tracked the location of cargo. These tools provided in-transit visibility and expedited deployment planning and execution. Yet, more could be done. In the need to arrive fast, one could eliminate the need to deploy equipment if the equipment was already waiting for the personnel in theater.[15]

The U.S. Army first tried that concept in Germany during the Cold War in 1969, as Prepositioning of Materiel Configured in Unit Sets (POMCUS) stocks allowed annual Reinforce Germany (REFORGER) exercises to have soldiers from the United States fall in on prepositioned equipment, which expedited the ability to deploy several divisions of personnel to Germany in the case of war with the Soviet Union. This concept inspired the Army and Navy to preposition ships with combat packages around the globe during the 1980s. The Navy established 13 Maritime Prepositioning Ships organized into three squadrons stationed in the Atlantic Ocean, Diego Garcia in the Indian Ocean, and Guam in the Pacific. Each squadron could equip and provide enough supplies to sustain a Marine Expeditionary Brigade for 30 days. With an increased focus on the Middle East, the Army likewise established its Army Prepositioning Stocks at Diego Garcia, which proved their value during Operation Desert Storm and later in Somalia. In 1981, the Navy also purchased from Sea Land Services eight container ships that could sail up to 33 knots, faster than any other container ship in the Maritime Sealife Command's inventory. By modifying them to have roll-on/roll-off capability, these ships served the Army well during Panama, Desert Shield/Desert Storm, and Operation Iraqi Freedom.[16]

One innovative concept in strategic deployment came with the deployment of the 3rd Armored Division to Saudi Arabia during Operation Desert Storm. In November 1990, the U.S. Third Army needed a heavy corps of two armored and one mechanized infantry division to shift from a defensive operation (Desert Shield) to an offensive one (Desert Storm). To reduce the competition for transportation assets, the 1st Infantry Division (Mechanized) would deploy from the United States, and the 2nd and 3rd Armored Divisions would deploy from Germany. The 2nd Armored Division had priority of rail and wheeled vehicle transportation, and the

---

15 Ibid.
16 Killblane, *Delivering Victory*, 155, 168.

3rd Armored Division feared that it would not arrive in time for the start of the ground offensive. There were no rail or heavy equipment transporters (HET) available, but the division staff learned that barges at the Mannheim River Terminal could transport tanks to the ports of Rotterdam and Antwerp since the 2nd Armored would ship out of Bremerhaven. Thus, the 3rd Armored Division set a precedent, and river barges became a common means of transporting combat units in Germany and arrived in time for the ground war.[17]

In five months, Central Command (CENTCOM) deployed two full corps of seven Army divisions (one airborne, one air assault, three armored, and two mechanized infantry) and one Marine division to Saudi Arabia in time for the ground offensive. This force overwhelmed the Iraq Army in 100 hours of combat. The United States had never deployed so many divisions so quickly since WWII. The next time that the United States went to war with Iraq, it would not have the luxury of time to deploy that many divisions.[18]

Anticipating that Saddam Hussein would not allow a slow buildup of divisions in Kuwait to prepare for Operation Iraqi Freedom, GEN Tommy Franks wanted to hit Baghdad from several directions at once: the 4th Infantry Division (Mechanized) from the north through Turkey, the 101st Air Assault Division from the west out of Kuwait, and the 3rd Infantry Division (Mechanized) and 1st Marine Division (Mechanized) from the south out of Kuwait. CENTCOM had long maintained a brigade in Kuwait for defensive purposes, rotated every six months. When it came time to replace one brigade of the 3rd Infantry Division (Mechanized) with another in October 2002, CENTCOM announced that it was also swapping out combat sets of equipment from the Army Preposition Stocks at Diego Garcia. In reality, it left two brigades of the 3rd Infantry Division in Kuwait. To complete the deployment, the division needed its third brigade and support package. The war would start two months after the deployment began in January 2003, although Turkey would not allow transit, delaying the 4th Infantry Division. Since the invasion of Cuba in 1898, the American Armed Forces had mastered the ability to deploy large masses of combat power around the globe.

---

17 Ibid., 185.
18 Killblane, *Delivering Victory*, 177-192.

Reception, Staging, Onward Movement, and Integration (RSO&I)

Strategic deployment ends when the units arrive at the ports of debarkation in the theater where the second logistical operation begins. Disembarking in theater may appear simple, but as learned during the Spanish American War, if not done right, it significantly delays all other operations. The offensive operation cannot begin until everything is in place. Like strategic deployment reception, staging, onward movement, and integration (RSO&I) have a transportation-centric focus except for integration. Because the ports of debarkations create a funnel effect in the deployment process, the goal has always been to expedite the process of offloading personnel, their equipment, and supplies and clearing the ports as rapidly as possible. Key to this effort has been the improved containerization and materiel handling equipment.

Early containerization began with boxes and barrels moved by dollies or ropes pulled through block and tackle by humans or draft animals. During the industrial revolution, steam power replaced muscle power. The invention of the gasoline-powered internal combustion engine replaced steam at the end of the 19th century, which allowed cranes to be mobile. Cargo, unfortunately, was loaded into nets and unloaded by hand—a slow process. The wooden pallet created in the 1920s permitted laborers to stack boxes and barrels on a stable platform, which could move the contents quicker than handling individual items. The forklift truck moved palletized contents around the port more easily and quickly. The problem of breakage and pilferage inspired the U.S. Army to package contents in an eight-by-six-by-six-foot container express (CONEX) in the 1960s, thus heralding the container revolution. As forklifts became larger in the 1980s, the metal containers increased in length to 22 military vans (MILVAN) and further expedited the movement of cargo. The fielding of the KALMAR Rough Terrain Cargo Handler (RTCH) in 1994 allowed Army stevedores to pick up and move 40 containers quickly. However, nothing could move bulk cargo and containers as rapidly and efficiently as overhead gantry cranes, which were limited to permanent port facilities. Unfortunately, Army stevedores did not always have the luxury to offload cargo at an established port facility.[19]

So, the Army instead needed to develop a more efficient method of delivering cargo across a bare beach. This problem became apparent during

---

19 Killblane, *Delivering Victory*, 33 and 140-141.

the Spanish-American War. During the invasion of Cuba, the Fifth Corps landed at Daiquiri and then shifted the port of debarkation for supplies to Siboney because the road led to the military objective—Santiago. Cargo was carried ashore from barges and launches (lighters), which proved very inefficient. The solution to unloading cargo over a bare beach grew from a similar British experience at Gallipoli, Turkey, during World War I. The British added a hinged landing ramp to barges, resulting in the Landing Craft Mechanized (LCM)-3, which the U.S. Army adopted as the LCM-6. Andrew Higgins developed an even smaller landing craft, vehicles, personnel (LCVP). The Navy took the idea of the hinged ramp to make larger landing crafts, utility (LCU). If the landing craft dropped a ramp on dry land, it expedited the ability to discharge cargo. The U.S. Navy also developed a fleet of landing ships, dock (LSD), personnel (LSP), and tank (LST), which offloaded larger quantities of supplies, personnel, and equipment. They would need these during World War II. These vessels improved the Army's capability to offload cargo onto a beach.[20]

Anticipating that the Axis Forces would deny the Allies entry to the seaports in their occupied territory, the Allies planned on amphibious landings near the ports. After the initial landings, the Transportation Port Commands would take over unloading additional personnel and supplies from the Engineer Special Brigades. This concept initially became known as over-the-beach operations and later changed to logistics over the shore (LOTS) after the war. To increase the volume of discharge, the Navy lifted the size restriction on Army watercraft, and the Army adopted LCUs. Because of the limited number of deep draft ports in Vietnam, the Army employed five medium boat (LCM-8) companies and three heavy boat (LCU-1466) companies to deliver cargo up and down the rivers and along the coast. With each generation of landing craft, LCUs increased in size until the LCU 2000 series was fielded in 1988, and the USAT *John U. D. Page* of the 1960s gave way to the Logistical Support Vessels commissioned in 1987. During joint LOTS operations, the landing craft generally performed lighterage from ship to shore but had limitations.[21]

The amphibious landings along North Africa during WWII revealed the limitations of landing craft. They could not drop the ramp on dry shore if the beach gradient was too shallow or if they ran onto underwater sandbars. So, the Army needed a truck that could swim. The resulting DUKW was a

---

20 Ibid., 14 and 41; and King, *Spearhead of Logistics*, 83.
21 Killblane, *Delivering Victory*, 120, 122, and 158-159.

2½-ton truck chassis with a boat body that could transfer the drive train from six-wheel drive to a propellor. These amphibious vehicles did not have to stop at the beach to transfer cargo but could drive right up to the depot, thus expediting the delivery of cargo. These saw extensive service during WWII. The drawback to the design was that the DUKW handled like a truck in the water, so the next generation of amphibious vehicles in the 1950s became boats that could drive on land. The family of Lighter Amphibious Resupply Cargo (LARC) vehicles had a five-ton, 15-ton and 60-ton capacity, but they still traveled no faster than ten knots. These saw service during the Vietnam War. The Lighter Air Cushion Vehicle (LACV) became the third evolution of amphibious vehicles. Hauling 30 tons at 40 knots, the LACV-30 (first fielded in 1983) could deliver 60 tons of cargo in half the time as the LARC-LX. Unfortunately, they were maintenance intensive. The Army eventually rid itself of these aircushion vehicles and let its amphibious fleet pass into obsolescence by 2001. Until then the increasing size of landing craft and amphibians allowed greater efficiency in offloading cargo over a bare beach; however, no matter how efficient, it could never match the speed and volume of stevedores moving cargo through a permanent deep-draft port.[22]

Another important aspect of reception, staging, onward movement, and integration that began during the planning process was the balance of the arrival of the support forces to the supported forces. Since the operations staff developed the concept of operations based on information gathered from the intelligence section, the logisticians had to convince the planners of the proper ratio of deployment. While the operators wanted to get as much combat power on the ground, the logisticians had to convince them to bring in enough transportation to move them from the port of debarkation and then sustain them. A historical example from the Spanish-American War was that soldiers only six miles from Siboney, Cuba, were starving even though the beach had plenty of food, as Fifth Corps had not brought enough wagons and mules to supply the soldiers. So, the initial deployment should contain enough combat power to secure the port of debarkation so that enough logistics can arrive to enable the arrival of more combat power to expand the operation. Most units deploy with three days of supply, so the initial phase of RSO&I had to include the accumulation of critical supplies: provisions, fuel, ammunition, and medical supplies. Once logisticians fill crucial needs of supply and transportation, then additional

---

22 Killblane, *Delivering Victory*, 67 and 121.

classes of supply and units that provide service functions arrive. So, the success of RSO&I is measured in throughput—how many military units or tons of cargo can pass through that port or across the beach.[23]

Sustainment

When military historians actually do study logistics, they tend to focus on sustainment because of its direct influence on battles and campaigns. When done right, sustaining the force appears invisible to the successful outcome, but bad logistics becomes glaringly obvious when something goes wrong. World War I provides a good place to start the study of sustainment influence on the U.S. Army. The U.S. Army's modern concept of war and logistics began during WWI and evolved with each conflict. The following examples illustrate different ways that logistics shaped the campaigns.

Because a continuous line of trenches stretched from the English Channel to the Swiss Alps, WWI introduced the broad front concept of war. This logistical concept dominated American warfare throughout the 20th century, with the Vietnam War as the anomaly. As part of the broad front concept of war, the American Expeditionary Force under General John J. Pershing organized the rear into a communication zone (COMMZ) under the control of the Service of Supply. The COMMZ was divided into the base, intermediate, and advanced sections. The base section represented each port of debarkation. Since rail provided the primary means of hauling passengers and cargo to the front, it transferred men and materiel from the base section to the intermediate section and then to the advanced section. The combat units occupied the advanced section. Pershing wanted 90 days of supply on hand in theater with 45 days stored at the base section, 30 days at a depot in the intermediate section, and 15 days in the advanced section. This approach remained the standard days of supply through World War II. Another policy of Pershing's that shaped the American concept of logistics was that theater logistics remained under the control of the theater commander. The stateside logisticians' responsibility ended at the port of debarkation. The COMMZ concept ended with the Vietnam War.[24]

WWI introduced trucks as a replacement for animal-drawn wagons that moved at a pace of three miles an hour on a good road. Trucks not only increased the speed of logistics, but the fuel that fed the internal

---

23 King, *Spearhead of Logistics*, 83-84.
24 Killblane, *Delivering Victory*, 22.

combustion engine fit under the truck's bed instead of competing for cargo space. The bed of the trucks were dedicated to carrying troops or supplies meant for troops. The Army in France issued truck battalions (then called motor supply trains) to every level of command from division to army and theater reserve. Since it was a new toy, no one except the French Reserve Mallet, a special unit, knew how to employ them properly. When the American corps and armies received requests to move men and supplies, they issued trucks and never saw the trucks again. Toward the war's end, the logisticians realized they needed to manage the trucks as a consolidated asset under a motor command. In other words, the command would receive a transportation request and dispatch trucks to fulfill the request. This concept served the American Army from then on. The question came down to how to best employ trucks along the line of communication.[25]

During WWI, the American Expeditionary Forces (AEF) created a Transportation Service to manage ports and rail since it had not yet learned how trucks fit into the line of communication. During the next war, the War Department created the Transportation Corps in July 1942 to manage port, rail, and traffic management while the Quartermaster Corps retained control of truck regiments and battalions. Army stevedores would discharge the ships and move cargo to the supply depots to properly manage transportation assets along the line of communication during WWII. Trucks assigned to the base section would then move cargo from the port depots to the railhead for transport by rail to the advanced section Truck battalions at the advanced section would then push cargo from the railhead a short distance to the rear. This concept eliminated the need for an intermediate section, which resulted in "skip echelon logistics."

During WWII and after, logisticians experimented with various methods of employing trucks to deliver cargo. Due to rail and bridge damage, trucks from the base section often had to deliver cargo to the advanced section, requiring the loaded 2½-ton cargo trucks to conduct long haul missions—delivering the cargo to the destination and back. Long-haul missions also required truck stops that provided drivers with food, rest, and maintenance. After engineers opened the Ledo Road from India to China, American truck drivers dropped their loaded trucks off so other drivers could continue the route while they returned with empty trucks.

---

25 Ibid., 16; and Colonel Brainerd Taylor, M.T.C., U.S.A., "Cooperation of Military and Commercial Interests in Motor Transportation Under a National Coordinating Authority," *Highway Transportation*, February 1920, 14-16.

The introduction of tractors and trailers allowed the tractors to exchange loaded trailers at a trailer transfer point for empty trailers. This line haul operation allowed different truck battalions during the Cold War to move trailers along the line of communication while allowing the drivers to return with their tractors to their home station at night. WWII refined the concept of distribution that facilitated the U.S. Army from then on.

The two main functions of logistics that heavily influence campaigns are lines of communication and days of supply. Because of the need for sustainment, lines of communication dictate the avenues of approach, and if transportation can sustain the required days of supply, the advance continues unhindered by logistics. This study will examine how logistics affected the advance. Because of the magnitude, diversity, and scope of World War II, it provides a number of different examples of how these functions shaped the campaigns.

The North Africa and Mediterranean Campaigns illustrate how lines of communication shaped the advance of armies. The German bomber threat out of Malta and Italy threatened Allied sea lines of communication in the Mediterranean. So, in 1942, the Allies had to land at three beaches in Morocco to secure the port of Oran. From there, the Allies advanced along the coast because a railroad connected Oran with Bizerte, Tunisia. Bizerte offered a short leap to Sicily. Securing Sicily provided a short sea line of communication to the toe of the boot of Italy. From Naples, the Fifth Army advanced up Italy at the pace of repairing railroad bridges. Once the Allies reduced the German Luftwaffe threat in the Mediterranean, they could then conduct an amphibious landing near Marseille in Southern France. Essentially, seaports and short sea lines of communication defined the conduct of the North Africa and Mediterranean campaigns.[26]

More often throughout history, units at the front have not run out of supplies because of a shortage of supplies but instead for a shortage of transportation. The Northern France Campaign during 1944 perfectly illustrated that. To prevent Germans from reinforcing the Normandy beaches with tanks during the Allied landings, the U.S. Army Air Force destroyed all the bridges the French resistance could not. Consequently, the Army engineers had to rebuild those bridges to sustain the First and Third Armies' rapid advance across France. The slow repair of bridges and construction of the fuel pipeline required far more trucks than originally anticipated.[27]

---

26 Killblane, *Delivering Victory*, 57-79.
27 Ibid., 84-89.

The Third Army was an armor-heavy force that consumed more fuel than the other armies, making fuel the logistical center of gravity. So, in August 1944, the Chief of Motor Transportation COL Loren A. Ayers and MAJ Gordon K. Gravelle borrowed additional trucks from the Normandy Base Section and devised a long-haul method of one-way truck traffic out of the railhead at St Lo to the First and Third Armies. Even though the Red Ball Express increased the delivery of fuel, the Third Army still ran out and came to a halt on September 1. The Air Force even tried to deliver fuel by aircraft but could not fulfill the massive fuel consumption rate. Consequently, the Third Army would have to delay until it had built up sufficient days of supply to go on the offensive, only to halt again when that fuel ran out. To compound the problem, the Ninth Army arrived in Normandy on September 5 and joined the right flank of the Third, thus increasing the demand for fuel. As the railhead advanced across Northern France with the repair of bridges, other express routes extended the supply by trucks. The four American armies could not sustain a push into Germany until November, when the rail line extended far enough to increase fuel supply. However, another shortage of supply halted offensive operations.[28]

The Sixth Army Group closing the Colmar Pocket in late 1944 illustrates how armies can advance only with sufficient days of supply. The Southern Line of Communication (SOLOC), originating in the Continental Base Section at Marseille, functioned as it should. The damage to the railroad bridges was not as extensive as in Northern France so that rail transportation could keep pace with the advancing U.S. Seventh and Free French Armies. Trucks pushed cargo from the port to the railhead and augmented rail to deliver to the front. When the Sixth Army Group launched an offensive to close the Colmar Pocket on November 15, the American divisions reached the Rhine by November 25. They then expended their allocation of artillery ammunition repelling the German counteroffensive during December and January and had to wait until January 20 to go on the offensive again. It halted again from February 1 to 8 before the U.S. XXI and First French Corps could close the pocket. The problem was not from a lack of transportation but a shortage of artillery ammunition production in the United States. The Sixth Army Group had advanced as far as its artillery supply allowed it. Then, it went into an operational pause until it built up enough days of supply to launch another offensive. The Sixth Army Group experienced three operational pauses before finally driving all the

---

28 Killblane, *Delivering Victory*, 89-91.

Germans across the Rhine River. While road and rail dominated the lines of communication in the Mediterranean and European Theaters of Operation, this was not always the case in other theaters.[29]

The Southwest Pacific Area (SWPA) provided a unique problem and solution to wresting control of New Guinea from the Japanese Army. Instead of advancing through the dense jungle from one end of the island to the other, General Douglass McArthur decided to leapfrog along the northern coast from Milne Bay on the eastern tip to Hollandia on the western end. The difference between littoral and amphibious operations is that the landing force reset at the port instead of returning to the fleet to reset. Using sailing vessels from the Small Ships Section during the summer of 1942, the 32nd and 41st infantry divisions took turns landing outside the next major port and fought their way in. Once they had secured the port, the next division would then conduct an amphibious landing outside the next port town and repeat the process.

Consequently, littoral operations eliminated the need for lengthy ground lines of communication. Freight ships and Liberty ships delivered supplies and equipment directly to the ports. The selection of ports depended on the range of fighter aircraft. When the 2nd Engineer Special Brigade (ESB) arrived in theater with its landing crafts mechanized (LCM) during the summer of 1943, it assumed responsibility for landings of the recently activated Sixth Army. In April 1944, the 2nd ESB delivered the 41st Infantry Division at Hollandia, which finally made way for the invasion of the Philippines. In all, the 2nd ESB conducted 28 amphibious landings. Inspired by this success, McArthur would repeat attempts to bypass the enemy during the Korean War with his Inchon and Wonsan landings in 1950. Army landing craft made littoral operations feasible.[30]

The increasing size of aircraft made sustainment three dimensional by adding air lines of communication. The advent of air transports entering WWII allowed the U.S. Army Air Force to open air bridges in theater. During the siege of Bastogne, Belgium, in December 1944, the Air Force had enough C-47 transports to sustain the 101st Airborne Division until the 4th Armored Division broke through and ended the siege. But the most significant air line of communication during the war was the

---

29 Jeffrey J. Clarke and Robert Ross Smith, *The European Theater of Operations; Riviera to the Rhine*. (Washington, DC: Center of Military History, 1993), 533-560; and Roland G. Ruppenthal, *Logistical Support of the Armies, Volume II: September 1944-May 1945*. (Washington, DC: Cener of Military History), 250-251, 266, 270, 272, 442, 445, and 451.
30 Killblane, *Delivering Victory*, 39-44.

Fourteenth Air Force effort to supply the Chinese Nationalist Army out of India in the China-Berma-India Theater of Operation. It was the only line of communication available until ground combat in Burma opened up a ground line of communication to China, and the Engineers could construct the Ledo Road through the mountains. After the war, the new U.S. Air Force borrowed lessons from the aerial supply of China to sustain the German population of Berlin during the blockade by the Soviet Union from June 1948 to September 1949. Even as U.S. Air Force transports increased in size and cargo capacity, an air line of communication could not compete with the tonnage delivered by ships or trucks. It rarely sustained any force larger than a division. At most, the air line of communication would augment the ground line of communication.[31]

Each subsequent war offered unique lessons in sustainment operations. The Korean War provides a good study of how the length of the line of communication can shape campaigns. A short line of communication with one's supply base offers quicker resupply and reinforcement. Once hammered back across the Nakdong River, the three American infantry divisions and several Republic of Korean (ROK) divisions held firm around the port of Pusan. The North Koreans' supply line stretched far from the 38th Parallel, not only delaying the delivery of supplies but also making it vulnerable to an amphibious landing at Inchon in August 1950. Severing the North Korean supply line caused a retreat back across the 38th Parallel. When the Eighth Army and X Corps reached the Chinese border, they were as far away from their supply base at Inchon as the North Koreans had been outside the Pusan perimeter. When the Chinese intervened, they drove the United Nations forces back battle after battle until the Battle of Chipyong-ni in February 1951, COL Paul Freeman, Commander of the 23rd Infantry, reassured his soldiers that they could be reinforced because they were closer to their supply base than the Chinese were to theirs and he was right. Consequently, the Eighth Army stabilized its defense along the 38th Parallel. The veterans jokingly referred to this back and forth as Operation Yo-yo. While the Korean War resembled the advance up the Italian Peninsula, the next war would more closely resemble the New Guinea Campaign in regard to sustainment.[32]

The Vietnam War provided a different reason for shortening the lines of

---

31 Ibid., 50-56.
32 During an Officer Professional Development meeting in 1988, GEN Paul Freeman, honorary commander of the 9th Infantry Regiment, explained to the officers of the 9th Infantry why the 23rd Infantry held its ground at Chipyong-ni.

communication, as the communist enemy had conventional forces from North Vietnam and insurgent forces from South Vietnam. For the American and South Vietnamese counterinsurgency efforts, there was no front or rear. In a war without an enemy front, the operational area was divided into four Corps Tactical Zones—I CTZ in the north and descending numerically to IV CTZ in the south—with units operating out of base camps along major roads or rivers. Consequently, the supply lines were vulnerable to guerrilla ambushes.

Because the country was narrow and long with an extensive coastline, the best way to keep the lines of communication short was to establish additional deep draft ports along the coast at Newport, Cam Ranh Bay, Qui Nhon, Vung Ro, Vung Tau, Cat Lai, Da Nang and Nha Trang, and beach ramps at Dong Tam, Phan Rang, Can Tho, Hue-Phu Bai and Dong Ha. The three main depots were at Saigon (later Long Binh), Cam Rahn Bay, and Qui Nhon. These depots maintained 45 days of supply of ammunition, with 15 days at the ammunition supply points and no more than 30 days of supply of fuel in country. In addition, the 1st Logistics Command opened up logistics over the shore operations at Sau Hugynh to support the 101st Airborne Division in opening up the coastal highway (QL1) in 1967 and Wunder Beach near Quang Tri to support the 1st Cavalry Division in I CTZ during 1968.[33]

From the Spanish-American War onward, the U.S. Army owned a watercraft fleet. By the Vietnam War, this fleet comprised mostly of landing craft and amphibious vehicles, but this large fleet gave the Army the unique capability to deliver cargo essentially anywhere along the coast of Vietnam. Consequently, Army watercraft was in great demand. The Army deployed as many as three heavy boat and five medium boat companies as well as four amphibious companies to perform lighterage in Vietnam. Because of this capability, the vast majority of supplies traveled by these vessels and vehicles to the sub-ports for final delivery to the bases by truck.[34]

The communist threat inspired innovations in convoy security. When Military Assistance Command, Vietnam (MACV) escalated the ground war during the summer of 1965, the enemy harassed convoys with small arms fire and mines. On September 2, 1967, in a prelude to the Tet Offensive, a North Vietnamese Army company launched the first large-scale ambush intended to destroy an entire convoy returning from Pleiku along the

---

33 LTG Joseph M. Heiser, Jr. *Vietnam Studies: Logistics Support*. (Washington, DC: Department of the Army, 1974), 25, 73, and 107.
34 Stanton, *Vietnam Order of Battle*; and Killblane, *Delivering Victory*, 149-150.

highway (QL19) through the Central Highlands. Since the II Field Force could not secure the entire route, the truck companies of 8th Transportation Group at Qui Nhon built gun trucks as a form of immediate defense until the nearest quick reaction force could arrive. From November 11 until May 1968, the enemy conducted large-scale ambushes weekly and mining and harassment fires daily. Over the span of a year, the truck companies experimented with gun truck designs and best procedures and managed to keep QL19 open. In 1968, the enemy began large-scale ambushes along the other supply routes in Vietnam. The gun trucks proved successful, but the Army considered the war in Vietnam an anomaly to the broad front concept of war and forgot the lessons of convoy security so hard learned. It would have to relearn those lessons in Iraq in 2003. Until then, the Army engaged in two more large-scale combat operations.

In 1991, Operation Desert Storm closed out the 20th century as the last large-scale war employing the broad front concept of war, which had originated during WWI. The Iraqi Army had built a defense in depth behind a series of anti-tank ditches and berms along the Kuwaiti border with weaker divisions up front and the elite Republican Guard divisions in the rear. The commander, General Norman Schwartzkopf, wanted to attack them from the flank—in what he called a "Hail Mary" maneuver. To accomplish this movement in Saudi Arabia required the secret buildup of forward logistic bases with 29 days of food, 5.2 days of fuel, and 45 days of ammunition while simultaneously moving two corps to the left without the enemy observing this. Schwarzkopf planned to conduct a massive air campaign to distract and blind the Iraqis during this lateral shift. Schwarzkopf asked his senior logistician, Lieutenant General William G. Pagonis, Commander of the 22nd Support Command, how long this move would take. Pagonis' movement control planners assured him they could complete the maneuver in three weeks. After the bombing campaign began on January 16, everything was in place to conduct the ground offensive on February 15. With lightning speed, the coalition defeated the Iraqi Army in one hundred hours of ground combat.[35]

Unlike the WWI broad-front concept, the situation for Operation Iraqi Freedom (OIF) in 2003 allowed coalition forces to advance deep into enemy territory, bypassing resistance to strike at the Iraqi capital and seat of government Baghdad. Bottled water proved to be the most important logistical concern during OIF I. The original distribution plan called for

---
35 Killblane, *Delivering Victory*, 187-189.

the 3rd Infantry Division to drink ROWPU purified water carried in water trailers; however, the division left its water trailers at home station because the soldiers preferred bottled water. This choice required far more trucks to deliver bottled water than had been allocated. To compound the problem, the operational readiness of the trucks already in theater was 20 percent less than planned. So, there were not enough trucks to sustain the 3rd Infantry Division's drive. On March 19, the 3rd Infantry Division left its assembly area with seven days of supply of water and advanced north, paralleling Highway 8 with the 1st Marine Division on its right. This situation meant the 3rd Infantry Division would run out of water on March 25. Anticipating an operational pause, the Third Army planners planned for a bombing campaign to pound the Iraqis until the trucks of the 7th Transportation Group could deliver enough water to renew the advance to Baghdad on March 29. The 3rd Infantry Division reached the Baghdad International Airport on April 3 and two days later entered the city. The next supply issue came from enemy interdiction.[36]

The mountain of containers accumulated in Saudi Arabia during Operation Desert Shield/Desert Storm had inspired the Army to reduce the logistical footprint by adopting the civilian logistical concept of on-time delivery, where the delivery of supplies would arrive on time, thus eliminating the need for warehousing or large quantities of supplies sitting on the ground. Army doctrine of the time also reduced the days of supply to the forward-deployed corps to no more than ten. This approach led to the Quartermaster Corps eliminating depot supply companies. In spite of this, the theater logistics in Kuwait built a significant supply depot (referred to as the Theater Distribution Center), and the Corps Support Command (COSCOM) at Logistical Support Area Anaconda in Iraq built the Corps Distribution Center (CDC). The new doctrine was not working and about to prove dangerous.[37]

In May 2003, the sustainment operation matured into a hub and spoke distribution operation with theater-level trucks out of Kuwait delivering to the CDC, where COSCOM trucks then pushed the cargo to the four multi-national divisions MND-North, -West, -Central, and -South. Reducing the logistical footprint caused COSCOM to maintain a minimum reserve of seven days of supply of the critical classes of supply—fuel and ammunition.

---

36 Richard E. Killblane, "For the Want of a Bottle of Water" in *The Long Haul; Historical Case Studies of Sustainment in Large-Scale Combat Operations*, Keith R. Beurskens, editor. (Ft Leavenworth, KS: Army University Press, 2018), 165-179.
37 Killblane, *Delivering Victory*, 255 and 262.

Another part of reducing the logistical footprint was hiring contractors to perform most of the duties of support, such as driving trucks. By 2004, civilian contractors drove four out of five trucks in American convoys.[38]

On Thursday night, April 8, 2004, the insurgents dropped eight bridges and overpasses around Convoy Support Center Scania, thus severing northbound traffic into the Sunni Triangle around Baghdad. The logistical clock was now ticking—the maneuver units were consuming their reserve supplies. The next day, the Madhi Militia ambushed any and every convoy trying to get in or out of Baghdad International Airport (BIAP), including the first large complex ambushes of the war (similar to September 2, 1968, in Vietnam).[39] The death of eight Kellog, Brown, and Root (KBR) drivers in an ambush caused contract drivers to quit in the hundreds. The next day, the movement control battalion coded the roads black, meaning any convoy venturing out faced imminent attack. By Easter Sunday, April 11, 2004, due to heavy fighting, the 1st Cavalry Division was two days from mission failure in fuel and ammunition. At the same time, the Madhi Militia attacked the southwest wall at BIAP, near where the trucks were parked. The enemy knew the soft underbelly of the 1st Cavalry Division was its dependency on trucks that delivered everything it needed to fight.

Brigadier General James Chambers, Commander of 13th COSCOM, employed emergency measures to deliver fuel and ammunition to the maneuver battalions of the III Corps. He directed any soldier with a driver's license to replace the KBR drivers, had tanks and Stryker vehicles escort the fuel convoys out of Scania, and had ammunition flown directly to BIAP for further distribution. After the April uprising, Chambers instituted a few changes. He opened two new lines of communication, one out of Turkey and a third out of Jordan, so Multi-National Corps-Iraq was no longer dependent upon a single line of communication from Kuwait. He also increased the days of supply at Anaconda to 15 and seven at the other hubs. Al Sadr's weekend uprising revealed the fallacy of on-time delivery and reducing the logistical footprint. In the commercial logistics that inspired it, no one was ambushing FEDEX trucks. There was a viable reason for stockpiling supplies in a war. Pershing was still right.[40]

---

38 Killblane, *Delivering Victory*, 260-265.
39 During OIF, the Army coined the terms simple and complex ambushes to denote the differences in size and scale. Simple ambushes usually comprised five-seven attackers firing small arms while a complex ambush comprised up to thirty attackers employing multiple weapon systems.
40 Killblane, *Delivering Victory*, 267-269.

Retrograde

Once the battle and war has been won, the interest in logistics dies very quickly. Studying how armies withdraw from theater is as interesting as watching soldiers try to turn their field gear into the central issue facility (CIF). Even logisticians find retrograde boring, but since Desert Storm, the United States has had to leave the theater the way it found it—no easy task. In regard to logistics operations, the U.S. Army defines retrograde as the withdrawal of units and equipment from the theater of operation. After the end of WWII, the focus of retrograde was sending military personnel home and leaving everything else in country, as it proved cheaper to sell or leave the supplies and equipment in place than to transport them back to the United States. In addition, much of the equipment was left to arm our allies. So, units generally waited their turn to retrograde depending upon the availability of strategic transportation.

The end of the Vietnam War was the first large-scale war in which the Army intended to retrograde personnel and equipment from the theater of operation. President Richard M. Nixon ran on a political campaign to pull Americans out of the Vietnam War. The incremental withdrawal began in June 1969, known as Operation Keystone, with each phase named after a bird. Keystone Eagle redeployed the 9th Infantry Division and 3rd Marine Division. The original plan had the units redeploying with all their equipment, but after the redeployment of the first brigade of the 9th ID, the decision was to turn the equipment over to the Army of the Republic of Vietnam (ARVN). Since the ARVN only wanted equipment fully operational, the 1st Logistics Command had to ship damaged or inoperable jeeps to rebuild on Taiwan, trucks to Army depot maintenance on Okinawa, and tanks and armored personnel carriers to the United States. When it came time to redeploy a brigade or division, the U.S. Military Advisory Command, Vietnam (MACV) assigned the personnel who had not completed their one-year tours to other units and zeroed out the units. It achieved the reductions through regular individual rotations by not replacing them from the United States. The last units returned during the spring of 1973. A small number of American servicemen and women remained in Vietnam until the fall of Saigon in April 1975, but they were not in combat roles. After Vietnam, the U.S. Army would have to retrograde everything it brought with it to war and retrograde would evolve into a fine art.[41]

---

41 Killblane, *Delivering Victory*, 151-152.

For the next large-scale war, it had taken five months to deploy the seven Army divisions and one Marine division of the Third Army to Saudi Arabia during Operation Desert Shield/Desert Storm. Fortunately, the units had not remained there long enough to accumulate excess equipment, so it only took seven months to redeploy them and the mountains of supplies. Since the 3rd Armored Division was scheduled for inactivation, the Army decided to leave its equipment and 250,000 tons of ammunition in Kuwait for any armored or mechanized brigade that reinforced Kuwait. So, retrograde from Saudi Arabia was essentially RSO&I and strategic deployment in reverse. The next two long wars would seriously challenge logisticians.[42]

The retrogrades from Iraq in 2011 and Afghanistan in 2014 provided immense challenges. In both cases, the Army had accumulated over a decade of materiel and excess equipment from the military and contractors. To compound matters, the heads of state agreed on withdrawal dates without the logisticians' consultation about the feasibility of retrograding everything by those dates. Political agreements required the Americans to leave nothing behind. Because of the political sensitivity, the theater commands in both countries created special organizations with sole responsibility for retrograde. These, in turn, created an alphabet of smaller teams with specific functions in retrograde. First, the teams had to identify what and how much needed to move. They then had to determine what to do with it. Some accumulated supplies and equipment could be entered into the supply system and reissued to other units. Non-mission essential equipment could be retrograded early. Next, the teams had to determine how to move it. The theater command developed the sequence of when each unit would retrograde. Iraq had 92 bases to close, so it collapsed the bases back to the main supply route and then closed the logistic bases from north to south. Fortunately, Iraq had a secure route to Kuwait. If time ran out, the Responsible Reset Task Force (R2TF) could push everything south to Kuwait and deal with it after the deadline.[43]

Afghanistan presented a more complex problem. At the height of the surge in 2009, 53,000 American servicemen and women served in Afghanistan. As a landlocked country, CENTCOM Materiel Retrograde Element (CRME) could not just push everything to a neighboring country. Unlike Iraq, the International Security Assistance Force (ISAF) units

---

42 Ibid., 190.
43 Killblane, *Delivering Victory*, 274-278.

retrograding were still heavily engaged in combat. The Afghan National Army would take only so much American equipment; the rest had to go. Since there was not enough time to retrograde everything, Army Materiel Command designed a machine that ate armored cars and turned them into scrap metal to sell to locals. The retrograde in Afghanistan reversed the expansion process by closing down bases in Regional Command-South, Regional Command-Southwest, and then Regional Command-East. Contract trucks transported most of the equipment along the Pakistan Ground Line of Communication (PakGLOC) to the port of Karachi. When Pakistan closed that port in 2013, the only way to retrograde equipment was by airlift. The Air Force loaded as much equipment on any plane flying to a friendly neighbor and deposited it there until intra-transportation could later deliver it to Kuwait. The logisticians jokingly called it Operation "Anywhere But Here." Fortunately, the new Afghan president, Ashraf Ghani, permitted 12,500 American troops to remain in Afghanistan after the December 31, 2014 deadline to train and assist the Afghan Army. Eventually, the number was reduced to 9,800, and the logisticians still had to meet the deadline. The logisticians in Iraq and Afghanistan had accomplished a monumental feat.[44]

Conclusion

Battles and campaigns represent only a limited study of war but have proven to be the most interesting. A more thorough study of war would encompass all aspects affecting the outcome. However, without understanding logistics, campaigns and battles lose much of their context and the study is confined to a small window of maneuver. The conduct of the military operation is primarily the product of the first three logistical operations. Strategic deployment and RSO&I are responsible for how much combat power one side can have and how quickly that combat power can be committed to the fight. During sustainment operations, lines of communication dictate the routes of advance while the days of supply support or hinder the rates of advance of the maneuver force regardless of their skill or fighting spirit. Those maneuver units must operate within the reach of the modes of transportation that supply them. The three modes of transportation (land, water, and air) allow a creative art to logistics in selecting the best lines of communication and alternatives. Service during sustainment has long-term

---

44 Killblane, *Delivering Victory*, 242-245.

effects on soldiers' morale and equipment maintenance. Continued low morale can lead to low esprit and poor fighting spirit. A lower operational readiness rate can also reduce the number of vehicles and vessels available for distribution and even combat vehicles for the fight. Taking these factors into account provides a better understanding of the outcome. Most students of war, and even logisticians, often ignore the last operation, retrograde, since it generally occurs after the war has concluded, but the war is not finished until everyone returns home. Regardless of the type of military contingency, from combat to disaster relief, logistics operations remain the same—only the consequences change. The study of logistics provides a more comprehensive understanding of war.

PART II

PRESENT SUSTAINING CONTEMPORARY WAR

6

AN ARMOR OFFICER'S PERSPECTIVE ON LOGISTICAL LITERACY

Rich Creed

Something that distinguished competent combat arms officers commissioned before the Army's reorganization into a modular force during the wars in Iraq and Afghanistan was their responsibility to understand, appreciate, and plan logistics. Most battalions were what we called 'pure,' meaning staffed from top to bottom with officers from a single branch except for the intelligence and chemical officers. Practically speaking, this meant that the battalion logistics and maintenance officers were not logisticians, and neither were all the logistics-focused duty positions in the headquarters company. This forced officers to learn all manifestations of tactical logistics, including the Army's maintenance and supply systems that enabled the flow of spare parts and supply commodities necessary for operations. One did not want to look ignorant when dealing with the logisticians in the forward support battalion, and more importantly, one did not want to be the one that caused your unit to fail because you could not support the hundreds of people depending on you.

As an Army, we may have inadvertently assumed risk once we went down the path of no longer making officers who were not logisticians responsible for logistics in tactical-level units. There was a genuine career-long benefit to learning the challenges of all things logistics firsthand as a company grade officer. Later in one's career, when one's broader responsibilities as a more senior leader demanded an understanding of logistical considerations during planning, those of us with practical experience could quickly communicate realistic expectations to the logistics and medical staff officers trying to support a course of action. I never got over the notion that it seemed unfair to hold the logistics community responsible for figuring out how to support plans drawn up by folks without appreciating the logistical intricacies of delivering results on the ground.

My experiences, good and bad, illustrate the intrinsic values of the old system and the importance of educating and training all officers in logistics to a greater degree than we seem to be doing today. It is unlikely that one can effectively lead large formations without some level of personal experience with logistics, since it is difficult to appreciate all the tangible and intangible factors of something if you have never done some aspect of that thing yourself. If nothing else, I hope to communicate the importance of a combat arms culture that considers logistics critical to professional development and perhaps elicit a few laughs.

Foundations

As a brand-new platoon leader, I had the most competent and demanding company commander in 1st Battalion, 69th Armor. CPT B. was probably the best company commander in the brigade, and since it was his second consecutive tour in Kitzingen, very few officers understood how things worked better than he did in that part of Germany in 1990. There was certainly nobody who had served in more duty positions within the battalion, and he had strong opinions about which jobs were most important to one's development as an armor officer. He made it pretty clear over the months I was a tank platoon leader that the best officers got the hardest jobs, and the hardest jobs were those that required tankers who understood logistics. So, while most lieutenants wanted to be scout or mortar platoon leaders when they left their tank platoons, he taught us that company executive officers and the support platoon leader were where real development took place. He had been a support platoon leader, HHC XO, and the battalion maintenance officer—all jobs that contribute to a mindset about logistics enabling maneuver. The way he developed us reflected that mindset, and it started on day one with how he coached his company XO and us platoon leaders. Nobody had to tell us he was serious, as everyone in the battalion knew he had relieved a platoon leader and company XO in the months before we arrived.

The garrison life of a tank platoon leader revolved around gunnery training, maintenance, and supply accountability. At least it did in C company—as evidenced by the monthly counseling statements we received—on yellow legal size 11x14 paper. The legal pad approach could be considered portentous if you thought about the legal implications of not following your commander's orders, something that did not occur to me at the time but made me think when I found the counselings in a folder several years ago. The good news was that the commander was completely transparent about us paying attention to gunnery, maintenance, and supply—and we were graded on those topics every month, no matter what else was going on. Since our focus here is logistics, I will not mention gunnery again other than to say that without logistics, there is no gunnery. As First Sergeant H. told us early and often, "Without fuel and spare parts, a tank is just a $4 million dollar pillbox."

One unbreakable practice imparted to us was that you topped off your very thirsty Abrams tanks at every opportunity and NEVER passed on a chance to get more fuel. EVER. This practice was because the first rule of tanking was you cannot run out of gas. Your platoon can never run out of gas, and your own tank had better never run out. Were that to happen, the minimum penalty was a letter of concern that was a prelude to being relieved of your platoon leader duties. Smart platoon leaders figured out that they needed to know the fuel status in each of their tanks all the time, and they mentally computed how much time they could run their engines before the next refueling when battalion pushed the logistics package to the company. When the fuelers arrived, it was a rapid drill to top off because pumping 300+ gallons into all 10-14 tanks needed to happen in about 45 minutes, often in the dark. That meant the whole crew, officers included, participated. In other words, if you wanted to be considered a tanker and not a tank-rider, you held a nozzle and pumped gas.

None of this was particularly difficult, but there were ways for things to go wrong. Learning how those things can go wrong was the beginning of achieving a professional appreciation for how friction at lower levels bubbles up to higher levels. 2LT Creed thought he had the fuel thing figured out until one weekday afternoon on a German public road during what passed for rush hour outside of Schweinfurt. We had spent the day maneuvering around the countryside and were heading back to our assembly area to refuel and eat. As we sat in traffic, my driver told me we were just about out of fuel, which seemed impossible. I asked the driver what the problem was. He said he thought the fuel transfer pump that moved fuel from the 200-odd gallon reserve cell to the main fuel cell was not working. I did not finish the @#$!@ in my response before the engine cut off 100 meters from an intersection with three M1A1 tanks and dozens of German civilian cars behind me.

Calling Captain B. on the radio to tell him my tank was stuck was more fun than telling him why, since no matter how you sliced it, the reason was "I'm out of gas." That I had gas I could not use was not particularly relevant to the conversation since I should have known that my fuel transfer pump was broken and made allowances. That I was blocking civilian traffic at the end of the day and would shortly be joined by the German police to direct traffic around our little convoy of road-bound behemoths added to the embarrassment, as did waiting for a fueler to show up once the traffic cleared, led by the first sergeant. When I got back, CPT B. did not yell or act angry. He simply told me, in a calm and somewhat ominous tone, to have maintenance confirm the fuel transfer problem as new and that my reserve fuel cell was full. I would have preferred a full-blown chewing out and sweat bullets until the maintenance team chief confirmed my story. Lesson learned. For the remainder of the time I served in heavy units, I always made sure to know how many tanks could not transfer fuel, because any planning had to account for fueling those vehicles more often than the rest. This practice included when I was a division operations officer and directed our staff to track those numbers during exercises, because there were times when they added up to companies of vehicles. Logistical details mattered.

In a Little Deeper

CPT B. moved on to take command of the headquarters and headquarters company, which in those days was responsible for executing all logistical support inside armor battalions like ours. My friend 1LT Dave S., who had been our company XO, was made the Battalion S4 (Logistics Officer) after leading the support platoon, which was a logical progression. There was a lot of gossip amongst the platoon leaders across the battalion about who would become the scout and mortar platoon leaders. I thought I had a shot until CPT B. told me I was going to be the support platoon leader to replace Dave. It turned out to be a pivotal experience in terms of appreciating the importance of logistics and the people who make it work at each echelon.

The old support platoon essentially did what a forward support company does now, except that soldiers were all enlisted armor soldiers other than the two senior NCOs. It had 16 2500-gallon HEMMT fuelers (that you never loaded fuller than 2250 gallons for reasons I cannot remember now), 14 HEMMT cargo trucks that hauled ammunition, six five-ton trucks that pulled field kitchen and water trailers, two HUMMWVs for me and my platoon sergeant, and over 70 soldiers sent to the support platoon because their tank company first sergeants found them expendable. If I had known then what I know now, we would have called the support platoon "The Dirty Dozens," but I digress. The support platoon leader was the account holder for all fuel, ammunition, and food in the battalion and was responsible for managing the battalion mess hall as well. There were essentially two "no-fail" responsibilities for the support platoon—one was accountability of resources tied to the budget, and the other was execution of support to the entire battalion in both garrison and the field. Depending upon what was going on and where, either responsibility could earn you unwanted attention. Nobody comes by to see the support platoon leader or his bosses, the S4 and HHC Commander, with good news and compliments for a job well done.

One exercise at Hohenfels in January 1991 stands out in terms of establishing my enduring attitude towards logistics. The two 3rd Infantry Division brigades that did not deploy to Operation Desert Shield were training individual ready reservists mobilized as casualty replacements while exercising against each other over a month. In that era, two brigades deployed to the training center and took turns as the 'blue' (training) and 'red' (opposing or enemy) force. It was not a big place by training center standards, so things could get very crowded in the training area, known as "the box." One way to mitigate the crowding was to put one brigade support area off the installation at a doctrinal distance from the maneuver units, usually about 15 or 20 kilometers away. Each of the battalion field trains was dispersed in different areas based on what the German authorities let us use, and we supported operations in the box using the German roads to and from the installation. The logistics packages (LOGPACs) we pushed to the maneuver units typically occurred after dark, which in January was around 5 pm. Doing it once a day was a good day's work, but because the prepackaged meals (MREs) in theater were strictly rationed due to the requirements of an impending war against Iraq, we were required to push two LOGPACs with fresh hot meals every day. Since it was the coldest and snowiest winter in Germany since 1940, we also pushed fuel twice a day because everyone idled their tanks and Bradleys more to stay warm and keep engine batteries charged.

It snowed almost 40 inches throughout the exercise, and our 8x8 trucks did not have snow chains. Since they were 8x8 trucks, we figured we would be okay, and while we were sort of right then, the results would probably raise some eyebrows today. The risk we accepted, because there were no alternatives, resulted in no injuries but a significant amount of minor damage to almost every vehicle. We viewed the mission as 'no fail,' since if we did not deliver, nobody would eat and the vehicles in the box would be out of gas. Canceling the training exercise was not an option despite the atrocious weather—a large part of the Army was preparing to fight a real war and we might be joining them in a couple months.

I learned a lot of interesting things, like how far to back a HEMMT fueler up so that it had enough speed to make it up a steep, snow-covered tank trail without sliding backward down the hill and colliding with the next truck in line. The answer we figured out after a set of broken headlights, a set of broken taillights, and smashed rear doors on the lead truck was about 150 meters. We also found out that the best move was to take the smashed rear doors off all the trucks that collided with each other over the exercise. I learned that drivers operating on less than four hours of sleep could knock down every 100-meter marker along 10 kilometers of civilian roads without once being involved in an actual accident or causing the Polizei to be called.

The exercise taught me that a HEMMT fueler can continue forward progress in a trout pond for about 50 meters or until the water is six feet deep. I learned that same night, after getting back to the release point with only one of the 20 trucks supposed to be behind me, that standard operating procedures can get overwhelmed by novel sights, like seeing a fuel tanker drive into a snow-covered trout pond. When asked, everyone could recite the SOP (standard operating procedure) that we keep moving no matter what, but nobody believed it applied during training when something weird happened. There were many other crazy things that happened that month, and the number one takeaway for this young armor officer trying to support a battalion logistically was this: logistics is a no-fail endeavor because the consequences are always real-world consequences. You cannot 'game it' in the real world. The other important thing I learned is that logistics leadership is no different than any other—provide intent and support your subordinates as they do their best. It still amazes me that damaging 26 of 31 of my trucks in a variety of ways never got me into any trouble at all, unlike what happened when a few of my hooligans cleaned enough salt and mud off a fueler to make it look just like a Budweiser® delivery truck.

For the next 20 years I never forgot how important training, small unit leadership, and accepting a little risk was to develop a mindset that accounted for the challenges of tactical logistics. In particular, the experiences made me think about what was possible under what conditions, and the risk associated with being overly optimistic about the likelihood of things going wrong.

Figuring It All Out, Kinda

A year later, I found myself in the Headquarters Company XO in 3rd Battalion, 37th Armor, working for another great leader tasked to fix a unit that had lost its way. CPT L. decided to focus on the company's internal supply problems and told me to fix the maintenance program. While I had been a tank company XO and understood the maintenance system, it was very different dealing with 212 vehicles and trailers than the 20 vehicles in a tank company. For one thing, there were a lot more variables: the variety of vehicles, lack of dedicated crews with the responsibility for individual vehicles, a more extensive fleet with specialized mechanics, and a diverse group of soldier specialties for whom maintenance was usually an afterthought.

Once again, I was blessed to have more senior Armor officers steeped in a culture that understood logistics in general and, in this case, maintenance specifically. What the battalion commander, battalion XO, and HHC commander needed was the focused application of relentless energy to help them re-establish a maintenance culture. Like almost every problem in the Army, maintenance needed engaged and innovative leadership to fix. This was the job that reinforced the notion that there are plenty of experts out there to help solve problems if you knew what questions to ask and provided the support to implement the suggestions they offer. That meant understanding how the system worked on a larger scale, making a case for keeping people in the motor pool doing the right things, and being allowed to enforce standards. It also meant occasionally upsetting some of your peers, your bosses, and sometimes your boss's boss. It was obvious in retrospect, but at the time not at all obvious, that when someone senior wanted things fixed and you start fixing them the only way that they can be fixed, the result might be some level of initial unhappiness about the cost. There is no easy way to make a maintenance program work without making a lot of people uninterested in maintenance do their jobs.

Two memories from that assignment really stuck. The first stemmed from a Bradley in our scout platoon that had a broken missile launcher that had not worked since hitting a mine during the First Gulf War two years earlier. It had been reported as broken—meaning that the vehicle was non-mission capable—for the longest period of any vehicle in the division. About a month into the job, I was surprised in one of the maintenance bays by a one-star general who asked me who oversaw maintenance in the battalion and when was that Bradley going to get fixed. I told the Deputy Division Commander for Support that I was responsible for company maintenance and that we had all our paperwork in order going back a long time. We were doing our best, but nobody knew what was wrong with the launcher or how to fix it. He asked if the missile technician from the support battalion had looked at it, and I said yes, but the technician could not figure it out. After looking at our paperwork and being satisfied that we (or at least I) were not the problem, he asked if we had a place for someone to sleep in the motor pool. When I told him we could make a spot, he told me that the missile technician, a warrant officer, would sleep in our motor pool until the launcher was fixed. My company commander quickly gave our battalion commander a heads-up when I called him five minutes after the general left. The warrant officer showed up with a duffle bag the next day, and the Bradley was repaired in less than two weeks. There is nothing like command emphasis to spark a little extra effort.

The other memory had to do, once again, with HEMMT fuelers. We had a shortage of heavy wheeled vehicle mechanics and HEMMT drivers who were not well trained in the maintenance of their trucks. Typically, it was diesel engine issues, but in this case, the failure to check fuel pumps garnered a lot more attention than any of us wanted. On a Friday afternoon during motor pool clean up, one of my junior mechanics came into my office and said, "Sir, there is a HEMMT leaking fuel." I asked if it was bad, and he said, "No," so I said, "Make sure there is a drip pan under it and go tell the support platoon leadership to get out there and fix it." About 15 minutes later, he came back and said he had the support platoon leadership out there, and they were trying to clean things up, but the leak was "worse." Being a little slow on the uptake, I said okay, let me know if they cannot handle it because I was busy trying to finish the week's paperwork and would be out in a little bit to check how things were going. Ten minutes later, he came running back in to tell me I needed to look because the leak was a lot worse. Having gotten my attention, I jogged out to see the last few hundred gallons of diesel pour out onto the concrete as dozens of Soldiers attempted to build a dam of dry sweep and sandbags around the truck. The main fuel pod itself had failed, not the vehicle fuel tank, as I had assumed.

This was not a small problem, even by 1993 standards, especially when our makeshift dam of sandbags and kitty litter burst, and a thousand-plus gallons of diesel fuel went running under the fence and into the ditch that surrounded the motor pool. I had seen fuel spills before, but nothing this big. The list of people to call and alert, in the pre-cell phone era, was daunting. None of them were happy, especially on a Friday afternoon. There were several lessons learned that day—the first of which was to be more curious about initial spot reports. The second was that not fixing small problems generally leads to much bigger problems, and when there are no 'extra' fuelers in a battalion, the loss of even one made a difference. It was the logistical equivalent of losing a tank platoon because it takes one fueler to top off four tanks during a typical day of operations. Understanding logistics in that light taught me to pay very close attention to the readiness level of fuelers in other assignments and as a planner. It taught me that there is also personal risk associated with logistics, as the Environmental Protection Agency has the authority to fine individuals for negligence, which did not happen in this case, thank goodness. On the lighter side, we all learned that a healthy dose of diesel in the ditches around the motor pool resulted in no mosquitoes for a couple of months that summer, so we had that going for us.

Take This Job

Several years later, as a tank company team commander in Bosnia, I had reasonable hopes of taking a second command of the battalion headquarters company the following year. One morning at Camp Bedrock, my battalion commander told me to join the brigade commander for breakfast in the mess hall. To my surprise, he was eating alone, so I grabbed a tray and joined him. After some small talk about how things were going, he said, "Rich, we've had seven brigade S4s in the past 18 months, and I need you to take the job. I want you to finish the deployment and assume the job this summer as you're the best option I have. Questions?" I wanted to say that I did not feel like the best option, did not want to be on brigade staff as a captain, and would be happy to stay in company commander another year. Or two. My real answer was, "Yes, sir. I'll do my best."

If previous assignments had not made me appreciate logistics and the risks associated with failure to approach sustainment seriously, this one certainly did. Immediately. The good thing about being a captain on a brigade staff is nobody expects you to be a genius, so you get a little time to figure things out. That was important because it quickly became obvious that there were many new things to figure out beyond the fundamentals I had learned about fuel, ammunition, and maintenance up to that point. For one thing, the brigade was a big organization to a company grade officer, and this one was spread over two different installations. The brigade S4 had a lot more responsibilities, one of which included planning and coordinating all movement over German roads and on the German railroad. We had overall responsibility for multiple dining facilities. We planned medical evacuation between the battalions and the medical company in the support battalion. Ultimately, we did much of the deployment planning when the brigade went to Kosovo, something that six deployments for gunnery and field training in southern Germany ultimately prepared us to do better than any lessons in a classroom could have done.

It took no time to figure out that brigade S4s are translators who communicate basic things between the logistics and maneuver communities in terms of requirements and the Army processes one uses to meet those requirements. It took a little longer to figure out that the S4 was also an interpreter who facilitates understanding between two communities with different perspectives and expectations about how logistics should work. Common words did not always resonate—for example, if the next operation involved an obstacle breaching operation in the context of a larger attack, it could not be assumed that the logisticians understood the timing that would affect refueling requirements. Likewise, one could not assume the maneuver units understood how long it takes to move heavily loaded big trucks along bad roads and trails if they were expected to push forward during that same attack. It was a big challenge to make sure the two communities were not talking past each other, they asked the right questions of each other, and I was value added to both sides as we tried to solve problems.

This Stuff is Hard

The daily realities of serving as an S4 for a couple of years forced one to learn new things almost every week because the expectations of the commander and everyone else on the staff were that you were the subject matter expert and the one to whom all logistics problems were sent for solutions. The first lesson was determining which problems were S4 problems to solve and which ones were unit problems to solve. Our support battalion was filled with experts responsible for executing sustainment across the brigade, and the other battalions had their own sustainment capabilities for which they were responsible. This situation meant that execution of sustainment was, for the most part, the responsibility of logisticians more senior than me who did not much appreciate getting advice and counsel from an armor captain.

The most spectacular public butt-chewing I have ever received was from one of the armor battalion commanders. I unwisely, and perhaps undiplomatically, told all the battalion executive officers (majors one and all) their logistics reporting status in an email and directed them to get their act together because division needed the data. It probably would have worked out fine if I had not, for some inexplicable reason, decided to include all the battalion commanders on the cc line. Airing every unit's dirty laundry in front of their commanders is generally a bad idea unless you are their commander. The dressing down waiting for me the next morning was a detailed, comprehensive, and quite brilliant recitation of the role of staff officers in our Army—of course, shared with the brigade executive officer, my rater. Besides learning a few etiquette lessons, I also learned that logistical shortfalls could cause professional embarrassment, incite strong feelings, and should be treated with a certain level of discretion in public forums when you are the junior person in the discussion. It was good to get this lesson out of the way during my second week on the job.

There were plenty of logistics experts in the support battalion and at division who could help an inexperienced brigade S4, and none of them were mind readers. Since it is difficult to know what one does not know, it was important to learn who those people were and what they did for the team. Just the act of doing that was an education in how sustainment was supposed to work, and it triggered a flood of questions simply because I realized that my job was to provide them with information, and they did the same for me. Working with higher and lower headquarters to sort out logistical issues was preparation for doing the same with other issues in the years ahead and provided insight about how nothing just happens because it was directed to happen on a piece of paper.

One enduring lesson the job taught was the importance of having a close relationship between planners and executors who know what they are talking about. One should not accept separation between what we once called combat arms, combat support, and combat service support—and the best practitioners of each need to be closely involved in planning. What this approach facilitated was the ability to advocate for someone else's requirements because they ultimately affected the entire combined arms team. Without teamwork, there is no combined arms.

So What

My personal experience with and an appreciation for logistics and sustainment overall made me a good division G3 because every plan or operation we published accounted for the logistics necessary to execute before we ever published orders. It was much easier to set realistic expectations for the brigades when one had experience planning their movement and sustainment. There was no 'us and them' when we addressed sustainment because that is not how the Army was ever intended to operate. Later, as a brigade commander with a brand-new captain as my S4, I was able to coach and teach someone trying their best instead of berating him or offering unhelpful advice to do better.' Likewise, it was easier to ensure our plans were executable because they accounted for logistical factors since I understood most of those factors from firsthand experience.

Perhaps most importantly, an understanding of logistics, even if not very current or particularly deep, helped ensure that during my tenure managing the Army doctrine program day to day we always account for the importance of logistics in our publications. With the Army again focused on large scale conventional warfare, logistics is not only one of the most critical things leaders need to understand, but it should also be the point of departure for those practicing operational art and crafting military strategy. One cannot have an appreciation for what Army formations can do without understanding the logistics that make them all work in the first place.

The U.S. Army fights in echelons of formations from platoons all the way up to theater armies. Each of those echelons requires sustainment support, and while the commodities and plans to deliver them come from the top down one needs to understand what happens in terms of execution from the bottom up. It matters because logistics that do not extend to the formations, vehicles, and people at the lower echelons so that they can function as intended leads to almost immediate failure before an operation even starts. Having a formation understanding of logistics, especially at the lower tactical echelons, is a critical part of understanding operational art. It is especially critical in the crucible of conventional warfare against a capable enemy, where the ability to sustain a plan in the most lethal and demanding environments devised by mankind can be the difference between success and failure.

Logistics-related assignments were foundational to my development as an Armor leader, and the things I learned as a support platoon leader,

HHC XO, and brigade were the toughest, most rewarding jobs I ever held. It is probably not a coincidence that so many past and current combat arms general officers had similar kinds of assignments, because they made us better leaders. We learned things from personal experience that our peers without these assignments could not. Hopefully, we will consider the benefits of exposing more maneuver officers to the world of logistics as we make decisions about the future of our Army organizations and how we organize them.

7

THE OPERATIONAL LEVEL OF WAR AND LOGISTICS

Kevin Benson

In the 21st century, the operational level of war is all about logistics—this became clear to me during my tour as the Chief of Strategic Plans for the Combined Forces Land Component Command (CFLCC) at the start of the Iraq War in 2003. In a theoretical world, planning proceeds in parallel with higher-level headquarters and in collaboration with subordinate headquarters. However, the planning for what would become Operation *Iraqi Freedom* did not occur in the theoretical world (an understatement to be sure). True planning at the level I served takes place in the space between military and policy requirements.

There must be a constant interaction between policy, strategy, and the operational level in war. Policymakers, in coordination with the Joint Chiefs of Staff, develop the policy goals for the use of force, which is the continuation of policy through other means. Planners at subordinate headquarters trust both the process and the Joint Staff support in representing what is needed to accomplish assigned tasks to the civilian policy making staff. Military leaders inform policymakers what strategic means or force can and, more importantly, cannot do. The operational art articulates how the ways and the means are used to link tactical actions and successes to attain the conditions needed to reach operational and strategic objectives. Commanders and staff at the strategic level of war sequence and sustain the force from the industrial base and depots over sea and air lines of communication to the theater of war. Commanders and staff at the operational level of warfare use operational art through campaign design and planning to sequence and sustain battles. Both levels of effort are about logistics and protection from physical and cyber effects.

Before the start of the war in Iraq in 2003, CENTCOM and Third Army—the Combined Force Land Component Command, or

CFLCC—were engaged with policymakers in refining the war plan and associated apportioned forces to ensure there was a sufficiently sized force of maneuver, protection, and logistics units on the ground before the decision to begin the war was taken. The following anecdote illustrates the level of effort involved in the discourse between military planners and policymakers.

In July 2002, I arrived at Third Army headquarters at Fort McPherson, Georgia. Shortly after settling into the plans section, I asked the lead planner how the planning was going for what we would do after we got to Baghdad, known as Phase IV. In the structure of the U.S. planning process Phase IV encompasses the anticipated tasks that are needed to complete meeting the requirements of military success. This means everything from coordinating the surrender of enemy forces to ensuring a level of security for UN, non-governmental relief organizations and other parties to assist in the reconstruction of a country. This was particularly important in writing a complete operations plan for Iraq. I was told the section was not working on this at all, and, at the moment, we were engaged in a task received through CENTCOM that came from the office of the Deputy Secretary of Defense. I asked the lead planner what the task was, and he handed me a copy of an interoffice memo that read, "We have a brigade on the ground. Why can't we go now?" The note was signed by the Deputy Secretary of Defense. I laughed. I felt this had to be an example of 'planner humor.'[1] After I finished chuckling, I said that while I appreciated a good joke, I was the colonel after all, and to please stop the joking and let me know what's actually going on. My lead planner just shook his head and assured me that this, in fact, was an actual question from the Office of the Deputy Secretary of Defense.

The task we received along with the note was to answer Wolfowitz's question as to why we could not begin the attack to Baghdad right then, in June 2002, as we had a brigade combat team on the ground in Kuwait. It was an amazing question and was answered only after we did the math on the amount of fuel, ammunition, water, and food it would take to move one M1A1 Abrams tank and its four-man crew from Kuwait to Baghdad, along with the support structure required to move that much fuel, ammunition,

---

[1] Planner humor covers the efforts made by all involved in the high-pressure environment of developing a plan that will work under the imperative of meeting deadlines, answering inquiries from higher headquarters, all of which have demands of receiving the required information "ten minutes ago." Planners engage in dark humor to lighten the atmosphere and sustain morale. Sometimes it even works.

water, and food. The pressure from Washington to address the task at hand had to be tempered by the reality of what was needed to support the effort and ensure a level of success.[2] The expectation in Washington was that with the application of air power and special operations forces—so successful in Afghanistan at that time—there would be similar results in Iraq. We had to convince policymakers in Washington that conditions in Iraq were different than in Afghanistan and there was more to the fight to get to Baghdad.

The planning effort required to ensure the balance of forces needed to fight AND sustain the anticipated battles we expected to fight during the attack to Baghdad, as well as the forces we needed to complete attaining the military conditions that met the policy objectives, was continuous. The effort of dealing with higher-level military staff and policy-making staff in Washington continued throughout the preparation for the opening of the war as well as during the march up to Baghdad. Again, these efforts always involved determining the correct force mix of maneuver, protection, and logistics forces. When Secretary of Defense Donald Rumsfeld decided not to execute a time phased force deployment list (TPFDL) and instead rely upon a process of requests for forces (RFFs), the work to refine the various force packages built around divisions, corps support units, and echelon above corps sustainment units took on a new level of urgency.[3]

The Secretary of Defense stated that since the president had not yet decided to go to war, Rumsfeld would not take one decision to begin the flow of forces that would commit the nation to war. The Secretary felt he could maintain control of the flow of forces by having Third Army/CFLCC and CENTCOM develop requests for forces. Reliance on the RFF process

---

2   This story is drawn from my book, *Expectation of Valor*. (Havertown, PA: Casemate Publishing, 2024), 4. Hereafter cited as *Expectation of Valor*.

3   The time phased force deployment list, TPFDL, is the product developed by planners from deploying units and U.S. Transportation Command. The development of a TPFDL involves precisely determining the amount of equipment and Soldiers deploying units have, the amount of shipping space needed to transport the equipment and Soldiers to a port in the theater of war, and then arranging the arrival times of ships and planes to ports in the U.S. so there is a smooth sequencing of forces and accompanying equipment, ensuring troops are there to link up with arriving equipment and in the order the ground commander requires. It is an enormously painstaking process. Units in the United States and those NOT in the theater of war all must GO TO the war. Given a finite number of transport ships and airplanes the sequencing of the effort is vital. Secretary Rumsfeld did not wish to make use of one TPFDL as he wanted to maintain tighter control over the number of units deploying for the operation. He required us to make use of the request for forces process, RFF. While the RFF made use of work done in preparation of the TPFDL, successfully getting a request for forces approved by the secretary was equally painstaking. The RFF also disrupted the established process for alerting, mobilizing and deploying Army Reserve and National Guard units.

required close coordination between the logistical and strategic planners to ensure that each force package within a specific RFF had the required mix of maneuver, protection, and logistics units.

During the weeks prior to the invasion in March 2003, we continued work on the flow of forces into the theater. Secretary Rumsfeld, through his military staff, continued to question CENTCOM and CFLCC about reducing the number of forces heading toward Iraq. Officers within the Office of the Secretary of Defense passed on the fact that the Secretary thought we were asking for too many forces. There was no force cap, but the number of troops in the RFF was too high for the Secretary to approve. A prime example of the urgency of this work follows.

On the evening of March 9, 2003, the CENTCOM Chief of Logistics, a major general, was personally involved in the analysis of force packages scheduled to arrive after the 3rd Cavalry. We did not know it at the time, but this work took place nine days before D-Day. This analysis was to determine what logistics units we would still need to have for Phase IV if a decision was taken not to flow the combat units associated with deployment order (DEPORD) 190. The combat units were 2nd Cavalry Regiment, 1st Armored Division, and 1st Cavalry Division. The analysis of the combat support (CS) and combat service support (CSS) requirements led us to conclude the units we needed to flow were primarily civil affairs, military police (regular and prison guard units), transportation (cargo and fuel), and engineers. With the reduction of combat units and other CS/CSS 'enablers' tied to those units, the total personnel requirement dropped from 60,148 to 11,594. We wanted to keep all units associated with the DEPORD flowing into the theater. In my assessment, I wrote we would at least require the 2nd Cavalry Regiment, but I recognized we might not have the choice.

We provided our assessment of DEPORD 190 CS/CSS units to our CFLCC senior leaders. This deployment order set out the units that would complete the service support structure necessary to sustain the combat units on the ground, what was necessary to sustain the combat units to Baghdad and beyond, and the combat units we felt were needed for Phase IV. We provided a number of 'Top 20' lists of active and reserve component units to the Office of the Secretary of Defense and the supporting staffs, including U.S. Army Forces Command (FORSCOM). It was essential to ensure that FORSCOM was fully and completely informed, given the rapidly changing nature of the units associated with the RFFs. The choice to build and refine RFFs based on unit deployment data rather than use a time phased force deployment list significantly upset an orderly alert and mobilization process of reserve forces.

Washington and the Pentagon were focused on the flow of forces as well as questioning the indispensability of just about every unit we felt we would need to continue the campaign. Every day we sent out another force flow chart based on the latest U.S. Transportation Command (TRANSCOM) slides to CENTCOM, which then forwarded our work to the Army and Joint Staff and, ultimately, to Secretary Rumsfeld's office. Our CFLCC focus remained on ensuring that every combat maneuver unit that arrived in theater was accompanied by the appropriate level of logistical and protection units.

The planning and plan refinement work we did required daily coordination with the logistics planners as well as the C4 himself and his key section leaders. The level of complexity of managing the deployment of forces increased daily. Our commanding general, Lieutenant General David McKiernan, kept us focused on maintaining the correct balance of maneuver, protection, and logistics forces needed to sequence and sustain the battles we envisioned en route to Baghdad.

We worked with our logistics staff section on a review of the 'Top 20' reserve component units from DEPORD 187—the order that completed the deployments of units associated with V Corps and CFLCC headquarters, as well as completing the flow of units in support of the 101st Airborne Division (Air Assault). The 'Top 20' review worked well. The FORSCOM commander and his staff worked with the National Guard Bureau and the Chief of the Army Reserve to ensure the 'Top 20' units were alerted, mobilized and deployed to Kuwait in support of the CFLCC commander and his scheme of maneuver for the invasion. We also received a significant number of calls for our priority for movement, especially given the constant calls from the Army staff, Joint staff, and the office of the Secretary of Defense all commands were receiving. MG Christianson our senior logistician and I, the senior planner personally led the review of our proposed sequence of flowing remaining Army forces into theater for Lt. Gen. McKiernan. The sequence was:

- Complete the closure of Force Module 1 (CFLCC and V Corps communications and logistics;
- Complete the closure of Force Modules 5, 6, and 7 (101st AASLT division with associated logistics units);
- Close Force Modules 2, 3, and 4 (3 Armored Cavalry Regiment (ACR) with associated logistics units) and then, in priority sequence from DEPORD 190:

○ Close Force Module 9 (2nd Cavalry Regiment),
○ Close Force Module 10 (1st Armored Division), and
○ Close Force Module 11 (1st Cavalry Division).

The plans team and I felt it was important to bring Force Modules 9, 10, and 11 from DEPORD 190 into Kuwait in sequence as we had designed and built the force modules to close in this priority.[4] We included CS/CSS capabilities, which not only supported the particular force module but also set the stage for the arrival and employment of the follow-on force modules. Changing the sequence always caused a last-minute review and redesign of individual force modules to ensure we had identified the correct units and capabilities. Any of these reviews, we felt, would cause a change in the total strategic lift required and, therefore, a change in the strategic transportation assets allocated by TRANSCOM.

CFLCC was not the only headquarters working with the force flow. During the pre-invasion period, we reviewed some TRANSCOM slides that showed a different view of force closure. TRANSCOM had to balance the sequencing of the ships, planes and trains needed to move and sustain forces. This TRANSCOM effort was also under scrutiny by the Secretary of Defense and his staff. CFLCC planners coordinated with the CENTCOM J5 on the topic of force closure/arrival. This is more evidence that even going to the war is an extension of policy, and under the direction of policy makers. The following is based on a CENTCOM J5 note to my planners and the Joint staff planners.

The slides that TRANSCOM prepared showed an adjusted picture of sequenced departure and arrival times in order to test the capability of the fleet, both ships and available passenger and cargo aircraft. TRANSCOM did not show the actual force flow because CFLCC and CENTCOM had not adjusted the force flow based on the new execution dates, as a result of the impact of RFF process. The CENTCOM planners resisted telling both CFLCC planners and the Joint staff to adjust our force arrival windows until more is known about the complete effect of RFFs on mobilization and movement. The TRANSCOM commander did not want to wait, so he

---

4 This sequenced arrival was necessary, we felt, for the beginning of Phase IV operations. The 2nd Cavalry was a wheeled force with plenty of fighting capability that would allow CFLCC to patrol our extended lines of communication. The arrival of the 1st Armored Division and 1st Cavalry Division would allow CFLCC to extend the necessary control of the country of Iraq and establish security conditions for UN, non-governmental and private volunteer organizations to operate. This would sustain the effort to attain the policy objectives of the U.S. government and in accord with UN Security Council resolutions.

adjusted the force arrival windows to show what was possible with the available sea and air shipping.

The CENTCOM planners wanted us to make sure our leadership knew this set of charts was a test excursion. CENTCOM planners assured us they would implement our priorities based on our adjusted desired arrival times for forces. The CENTCOM planners were struggling to make sure the CENTCOM commander understood this was just one excursion to test the TRANSCOM system. WE were directed to only pay attention to the final date for the closure of the entire force. That date was the only germane point and the focus of the test computer run.[5]

Throughout this period, indeed throughout the planning and execution effort associated with our campaign, I was constantly reminded of Moltke's dictum, which states that an error in the initial placement of forces could not be corrected over the course of a campaign. The struggle to bring forces into Kuwait and on to Iraq in the correct sequence and organized with the best mix of maneuver, protection and logistics forces was constant. An interruption of the force flow, from the breakdown of one ship to a delay in mobilization had to be accounted for in developing and monitoring the execution of the plan. For example, the use of the 4th Infantry Division in the operation.

There was consideration of using the 4th Infantry Division (4ID) to expand the battlefield and attack into Iraq from the north through Turkey. This concept required a detailed analysis of air and seaports in Turkey, the lines of communication from the seaports to the Turk-Iraq border, and the sustainment requirements of the force needed to maintain the lines of communication. In addition to requiring extensive collaboration with the U.S. Army Europe (USAREUR) commander and staff, the proposed use of this expansion of the Third Army/CFLCC concept of the operation depended upon the Departments of Defense and State to gain approval of the Turkish government. In the end, however, the Turkish government did not allow large conventional forces to use Turkish ports and attack into Iraq from the north. The late, but not unanticipated, decision by the Turkish government forced another logistical analysis to fit the entire 4ID force package into the flow of forces into the air and seaports of debarkation in Kuwait. The sequencing of the 4th and its equipment as well as what to do with the cargo ships carrying the 4th's equipment was another effort for planners and logisticians.

---

5   *Expectation of Valor*, 70.

Based on a question from the Pentagon there was a directive for Army staff, FORSCOM and CFLCC planners to consider the use of the cargo ships that moved the 4ID to embark another heavy division and position it off the coast of Turkey, 'in case' the Turkish government changed its mind and allowed U.S. forces to transit Turkey and enter Iraq from the north. During a coordinating video-teleconference between CFLCC, CENTCOM, and the Army staff, we learned of 'rhetoric' from within the Office of the Secretary of Defense was leading some to conclude Secretary Rumsfeld was too impatient to wait for the Turkish process to play out, and thus would not support another heavy force waiting for clearance into Turkey. The conduct of war and execution of plans is indeed an "extension of policy by other means."

The effort to redirect the 4ID clearly involved the commanding general of the division, Major General Raymond Odierno. He had concerns about the division moving to the south. Odierno had 40 people from the 4ID signal company on the ground in Turkey. We had to ensure these Soldiers and their vital digital equipment would be linked up with the division. USAREUR handled this transportation problem. Odierno wanted the 44th Signal Company, which was essential to his division's communications systems and the digital backbone of the division, established before the combat forces of the division arrived in Kuwait. The 4ID was the Army's sole 'digital' division. To ensure the full effect of the division's combat power was available for use in battle the sequencing, establishment and sustainment of the supporting divisional communications network was a vital first step in getting the division to Kuwait and then onward into Iraq.

From the USNS *Cape Horn*, Odierno wanted to ensure cargo-handling cranes were re-embarked for cargo transfer use in Kuwait. Odierno was deeply involved in the sequencing of the movement of his division. No detail was too small for him to ensure the 4ID arrived ready to enter the fight. This conference was long but necessary, as the 4ID was finally moving toward Kuwait for the fight in the main effort.

During all of the coordinating video teleconferences leading to the arrival of the 4ID the scheduling of passenger and cargo flow was discussed at length. I felt I needed to make the points that our movement planners were completely involved in the process, and scheduling would be managed by validating the arrival of passengers tied to the unloading of ships as they arrived in port. We learned that the 4ID advanced party, which had 516 troopers and a port support activity of another 200 with some pallets, would depart Fort Hood on or around March 25. Brigadier

General Speakes, a 4ID assistant division commander, would coordinate with the V Corps rear headquarters for space in one of the desert camps. The FORSCOM logistics chief coordinated with our C4, on the coordination of the arrival of the 4ID and the continued flow of force packages 186 and 187, elements needed to complete the 101st Airborne Division and V Corps support units. I assured all involved in the conference that we could accept the arrival of both sets of forces. The absolute need for the CFLCC J4 and J5 planners to work together was reinforced as at the time of the movement of the 4ID, CFLCC was also overseeing the V Corps and I Marine Expeditionary Force (MEF) attacks toward Baghdad. This effort highlights the need for the ability to sequence and sustain battles as a vital part of the operational level of war.

In addition to the planning for the arrival of forces into the theater and the attendant reception, staging, and onward movement of these forces, the operational and logistical planners were also under pressure to refine the plans for the departure of forces. CFLCC J4 and J5 planners continued work on the redeployment portion of our Phase IV plan, Eclipse II. As ironic as it felt at times, this was an important portion of Phase IV, as we had to coordinate the entry of forces into the theater and the departure of others.

The C5 Plans section focus ranged from preparations for the mission handover of CJTF-7 to V Corps, coordinating the integration of Coalition forces, monitoring the order of redeployment and handover of responsibilities to V Corps, and the establishment of bases and sustaining structures for the continuing mission, as well as retrograde and redeployment of units being relieved. Integrating Coalition forces into the overall Phase IV force truly required thinking through the logistical requirements of a multi-national force. For example, the Spanish government would commit a brigade task force. The caveat was clear: the Spanish would not consider that the Spanish area of operations should be anywhere but in the south of Iraq and within a one-hour round trip flight from the Spanish hospital ship offshore from Umm Qasr. The Spanish people were very sensitive to the conditions of the deployment, as Spanish politicians had told their people that the deployment was tied to the conduct of humanitarian tasks. Every nation that committed forces to the "coalition of the willing" arrived with unique sustainment requirements. Each national requirement demanded coordination between what the U.S. forces would provide and what was a national responsibility. Given the limited number of air and seaports, the sequencing of the arrival and sustainment efforts added a degree of difficulty to the planning and execution effort. The operational level of war is all about logistics and the sequencing and sustaining of battles. This is also true for stability and support operations. The discourse of the logisticians working group covered the expected size of deploying forces, who might need various types of support, and when CFLCC might expect various national elements to come to theater and conduct site reconnaissance.

Also indicative of the complexity of war, however, other events took up my time. As an example, on June 12, 2003, there was a snowflake[6] from Washington, which was yet another reminder that war is an extension of policy. We knew General Franks would depart for consultations in Washington that day. I reported to LTG McKiernan just prior to his scheduled departure that I had a fast request from the CENTCOM J3 to answer a snowflake from Secretary Rumsfeld. The Rumsfeld snowflake said:

I would like a projection on how we are going to be pulling down Kuwait and remaining Gulf states over the next three months...We have some 80,600 in Kuwait and another 15,800 in the remaining gulf states...I would think that at this point you might be able to give me a fix as to what your projection is between now and 1 Oct 03.

I was confident in my understanding of McKiernan's intent and did not have the opportunity to clear my response through him, based on the need for a rapid reply. I informed McKiernan that I replied:

Our retrograde from Kuwait is conditions based, the primary condition being a frank assessment of the security situation in Iraq by CJTF-7 and CPA and the ability of the forces to sustain that security. These forces are coalition and Iraqi. As we come to our series of decision points in the upcoming months if conditions allow, we will retrograde combat units and supporting combat support and combat service support units, divisional, EAD [echelon above division] and EAC [echelon above corps]. We will leave in place, in Kuwait and Iraq, a theater support structure appropriate to supporting the force that remains in Kuwait. This structure will meet both US and coalition support requirements (as needed). The drawdown of forces in Kuwait will be proportionate to the draw down of forces in Iraq.[7]

These events I cite highlight the continuous interaction of J4 and J5 planners to monitor the total effort needed to sequence and sustain

---

6 A 'snowflake' was the colloquial term for Secretary of Defense Rumsfeld's daily communication memoranda.
7 Expectation of Valor, 234. Combined Joint Task Force, CJTF-7, was the name of the headquarters that succeeded CFLCC in command of all U.S. and coalition forces in Iraq. The relief took place in mid-June 2003. The Coalition Provisional Authority, CPA, was the organization led by AMB Jerry Bremer. CPA directed the reestablishment of conditions for handing over control of the country to a new Iraqi government.

the battles fought during the execution of the opening of Operation *Iraqi Freedom* and *COBRA II*. War at the operational level demands the broadest understanding of the commanders' intent for both the operational level commander and the strategic level commander. The conduct of warfare at the operational level also demands a continuous running estimate and the agility to deal with friction.

The tension between ensuring there were enough forces in theater to begin and sustain the campaign also highlights the need to pay attention to the requirement to 'set the theater.' In the future, U.S. forces will continue to go *TO* the war; thus, the interrelationship of sustainment and maneuver will remain essential. It is also worth bearing in mind that the work done in Kuwait and Iraq in 2002/2003 to 'set the theater' was accomplished under conditions of air supremacy, so the extended CFLCC lines of communication were not seriously threatened until much later in the campaign, and then by ground unconventional or irregular forces. There was no threat from long-range fire systems or aerial fires.

As we have seen from observations from the Russia-Ukraine war, the Army must come to grips with the notion of 'contested logistics.' Given the increasing range of precision munitions and the widespread use of armed drones, we will need to consider the maneuver of sustainment forces as a part of the scheme of maneuver within the concept of the operation of a plan or order. We can no longer expect resupply convoys to drive along supply routes without these elements maneuvering underneath the layered protection of EW, space, and air defense/tactical ballistic missile defense. This includes individual ambulances and other trucks carrying out medical and casualty evacuation. Sustainment forces cannot simply organize resupply columns without extensive coordination with protection forces ranging from Military Police to Air Defense. The development of refined battlefield architecture must be an appreciation of protection forces throughout maneuver zones to ensure a high degree of coverage for supply routes and sustainment units.

Accepting the projections of engaging in combat with peer adversaries using conventional and unconventional forces, supply columns and sustainment forces must maneuver underneath layered protection of counter-air/drone/missile defenses as well as cyber protection. The terminal of a fuel pipeline, for example, will require a layered defense.

The central element of 21st century warfare will be a broadly shared common operational picture. At a minimum, all vehicles operating in the combat zone will be equipped with tracking/reporting devices to assure a reasonably accurate view of actions within the battle area. Thus, if all sustainment vehicles as well as fighting vehicles will operate with 21st century tracking/reporting devices. The broadly shared common operational picture then must be available to all headquarters from maneuver, protection and logistics for a total picture of the ongoing battle. The conduct of battle will demand incorporation of all systems aimed at sequencing and sustaining the battle leading to victory.

Warfare in the remainder of the 21st century will demand a thorough consideration of the sustainment of a campaign for the anticipated duration of the fight. The consideration of the campaign must include the movement of forces from the continental United States, from fort to air and seaport, and the use of sea and air lines of communication to the theater of war. Upon arrival in the theater, the details of reception, staging, onward movement, and integration of forces will stress commanders and staff. The reality of warfare in the remainder of the 21st century will demand general staff officers understand the interrelationship of maneuver, intelligence, protection and sustainment in the effort to plan and execute campaigns against peer adversaries.

The operational level of war is all about logistics. As armies rediscover the requirements for fighting a peer competitor under conditions of warfare that will include unconventional, conventional, and special operations forces, they must regain a deep understanding of the logistical underpinning the successful prosecution of 21st century campaigns will demand. From magazines, depots, and ports in home nations, over lengthy and contested lines of communication to theaters of war, the reality of "professionals study logistics" will truly come to the fore in staff and war colleges. Militaries must sequence and sustain campaigns, major operations, battles, and engagements at all levels of war. Indeed, while amateurs will continue to be enamored of tactics, wars are won at the operational and strategic levels. Success at the operational and strategic levels requires professionals to study logistics.

8

AT THE END OF A 6,000-MILE SCREWDRIVER

Francis Park

Afghanistan, the proverbial graveyard of empires, was the location of the longest war in the history of the United States and that of the North Atlantic Treaty Organization (NATO). The coalition built from NATO members and other countries that fought in Afghanistan comprised a sizeable force that was engaged in combat operations for well over a decade.[1] None of the combatants had planned prior to 2001 to embark on combat operations there, in a place far removed from home and one of the most difficult places in the world to sustain logistically. The examination of combat operations in Afghanistan lends itself to some observations about logistics, even with the return to more traditional forms of warfare since the end of the Afghanistan war.

One way to examine logistics as a whole is through operational art. Colonel Stephen Kidder, a former professor of military strategy and operations at the U.S. Army War College, offered a useful framework for examining operational art.[2] This framework has four elements: strategy, campaigning, force deployment, and sustainment.

In planning, strategy derives direction from policy and determines the objectives for employing the military instrument of military power and, in turn, the termination criteria for a military campaign. That strategy is realized through campaigning, which provides the detailed coordination

---

1 Although NATO convention is normally to refer to itself as an alliance, the addition of non-NATO troop contributing nations made the composition of the force in Afghanistan that of a coalition.
2 Prior to teaching at the Army War College, Kidder was a former chief of plans at U.S. Central Command. For a more detailed treatment of this model, see Francis J.H. Park, "The Unfulfilled Promise: The Development of Operational Art in the U.S. Military, 1973-1997" (doctoral dissertation, Lawrence, Kan., University of Kansas, 2012), chap. 1, https://kuscholarworks.ku.edu/handle/1808/10439.

(usually tactical) of capabilities and resources necessary to attain the ends of the strategy. The focus on implementation in campaigning provides a method for vetting the strategy.³

For that reason, the conduct of campaigning is beholden to the strategy that a campaign plan implements. The implementation of that strategic guidance is especially important in the relationship of campaign to force deployment and force sustainment. Decisions of strategy and campaigning involve the commitment of resources at the national and theater level, which requires careful planning prior to execution. Failures of campaigning, and by extension force deployment and sustainment, are not correctable through tactics alone, especially when fighting at expeditionary reach.

Force deployment not only includes the national strategic deployment of forces into a theater of war, but also the generation of those forces at home station incident to deployment. The challenges of deployment also exist for theater strategic movement of forces into the theater of operations and the operational maneuver of those forces within it. Force sustainment exists at much the same levels, whether national strategic from home station down to the 'last tactical mile,' especially in expeditionary warfare. Failures of sustainment at higher levels are not solely correctable by better execution at lower levels. Failures of strategic sustainment virtually guarantee hard times, if not culmination of the forward-deployed forces in a theater of operations.

As to its location, Afghanistan is a particularly unforgiving place to fight. The issue was operational reach, defined in American joint doctrine as "the distance and duration across which a joint force can successfully employ military capabilities."⁴ Even under the best circumstances, projecting and sustaining power at expeditionary distance over the full duration of a military campaign was a tough task.

As a land-locked country, Afghanistan had no seaports, and those in neighboring countries were in a hostile Iran or a reluctant Pakistan.

---

3   A necessary caveat to this particular analysis is that it does not treat the enemy's strategies to the same level of detail as those of the coalition. A more complete study would include a pairwise comparison of the coalition's strategy to those of its enemies. However, the enemy's strategy did not significantly change the demands placed on commanders and logisticians during the campaign. In an examination of the logistics of the Afghanistan campaign, the logistical problems and solutions were far more subject to the physical and political environment than the enemy threat.
4   Office of the Chairman of the Joint Chiefs of Staff, *DoD Dictionary of Military and Associated Terms, as of February 2023*. (Washington, D.C.: The Joint Staff, 2023), 50 and 146.

Even cargo unloading seaports in Pakistan had to traverse ground lines of communication (GLOC) in Baluchistan or the Federally Administered Tribal Areas and Northwest Frontier Province (Khyber Pakhtunkhwa since 2010) that were safe havens for the Taliban.

Airports were no better; any airports in the region that were available for sustainment were either remnants of the former Soviet Union (or its occupation) or had to be built. Once in Afghanistan proper, much of the terrain was impassable to heavy cargo trucks.

The main supply route within Afghanistan was Highway 1 (often called the Ring Road, as it is largely a path around the Hindu Kush mountains), the single north-south road connector between South Asia and Central Asia. Other spoke roads extended to the periphery, but in many cases, some smaller outlying bases were only sustainable through fixed-wing or rotary-wing airlift.

Military operations in Afghanistan were made even harder by the lack of any significant military posture in the region before the war. From a doctrinal standpoint, military posture comprises forces, footprints, and agreements, the latter two dependent on the access, basing, and overflight necessary to authorize coalition forces to operate within the theater of operations.[5] All would have to be built and sustained during the entire duration of the war.

The Coalition's Three Strategies from 2001 to 2014

The coalition's combat operations in Afghanistan reflected three basic strategies from 2001 to the end of 2014. Each of those strategies had distinctly different campaign approaches, manifesting in the deployment and sustainment of the forces in the theater of operations. From 2001 to 2003, the campaign was principally a punitive expedition in the immediate wake of the attacks on September 11, 2001. As the theater of operations matured from 2003 to 2009, the coalition transitioned to leading combat operations while starting efforts to build an Afghan security establishment. By 2009, the need to build the Afghan National Defense and Security Forces (ANDSF) became a critical element of population-centric counterinsurgency operations that went until the end of 2014 and the transition from Operation ENDURING FREEDOM (OEF) to Operation FREEDOM'S SENTINEL, the nominally

---

5   Joint Chiefs of Staff, *Joint Logistics*, JP 4-0. (Washington, D.C.: Joint Chiefs of Staff, 2023), IV–16, A-1.

'non-combat' train, advise, and assist mission.⁶ Deploying the forces against those strategies and their sustainment was as critical an effort as the strategies themselves, especially with the coalition's size at 50 countries at its peak.

2001-2003: Punitive Expedition While Setting the Theater

The attacks on the United States on September 11, 2001, generated an overwhelming outcry for a military response. As Sir Michael Howard observed shortly after the attacks, "It cried for immediate and spectacular vengeance to be inflicted by America's own armed forces."⁷ The resultant strategy was a punitive expedition to destroy al-Qaeda, the Taliban, and their allies in Afghanistan. Thus began the hunt for Osama bin Laden, Mullah Omar, and other al-Qaeda and Taliban senior leaders, the so-called "HVTs" (high value targets). What was conspicuously absent from the strategy was anything that looked like a permanent presence or nation-building.⁸

Moreover, Afghanistan was an austere theater, and theater strategic posture was required to enable forces within Afghanistan to operate. The initial deployment of forces involved the establishment of an intermediate staging base at Karshi-Khanabad (K2) Air Base, Uzbekistan.⁹ Pakistan provided initial permissions to establish a limited presence at four bases to support air operations from the Arabian Gulf.¹⁰ While there were seaports in Pakistan, none were employed initially, as most ground forces transited from K2. The exception was the 1st Marine Expeditionary Brigade, which deployed using organic rotary wing lift from amphibious assault ships to seize bases in Kandahar province. The scarcity of airports and the initial

---

6   While the ANDSF term distinguished between the military (defense) and police (security) forces, prior to 2015, the term for those forces were Afghan National Security Forces (ANSF), which did not make that formal distinction.
7   Carter Malkasian, *The American War in Afghanistan: A History.* (New York: Oxford University Press, 2021), 59; Michael Howard, "What's in A Name? How to Fight Terrorism," *Foreign Affairs*, December 1, 2001, https://www.foreignaffairs.com/united-states/whats-name-how-fight-terrorism.
8   Edmund J. Degen and Mark J. Reardon, eds., *Modern War in an Ancient Land: The United States Army in Afghanistan, 2001–2014, Volume 1.* (Washington: U.S. Army Center of Military History, 2021), 178.
9   Degen and Reardon, *Modern War in an Ancient Land: The United States Army in Afghanistan, 2001–2014, Volume 1,* 57.
10 Kamran Khan and Thomas E. Ricks, "U.S. Military Begins Shift from Bases in Pakistan," *Washington Post,* January 10, 2002, https://www.washingtonpost.com/archive/politics/2002/01/11/us-military-begins-shift-from-bases-in-pakistan/50a81af9-9739-41ad-9700-6392e8b6268e/.

absence of ground-based resupply increased the requirement for helicopters to move tactical forces. As a result, the first forces on the ground were U.S. special operations forces. Next were two incomplete brigade combat teams from the 10th Mountain Division and 101st Airborne Division and battalion-size British and Canadian contributions.

The theater of operations was so austere that sending the 82nd Airborne Division's ready brigade would have outstripped the capacity of the logistical infrastructure to sustain it.[11] Additionally, leaving behind artillery made the force heavily reliant on close air support provided from air bases on the Arabian Peninsula or launched from aircraft carriers in the Arabian Sea. The late 2001 establishment of forward operating bases at former Soviet air bases in Bagram and Kandahar helped slightly, as it provided logistical hubs in the theater rather than routing all support through K2, which closed to coalition use in 2005.

2003-2009: Coalition-Led Combat Operations

Establishing the logistical hubs at Bagram and Kandahar enabled the coalition to expand its footprint. After the conclusion of Operation ANACONDA, the environment appeared to be largely benign, and the threat posed by the Taliban was far less than during the first eight months of operations. In line with the reduced threat, the 2001 Bonn Conference led to establishing an Afghan Interim Authority in Kabul and a small peacekeeping force called the International Security Assistance Force (ISAF), whose activities were separate from the American-dominated OEF counterterrorism mission.

ISAF's tasks included military reform; judicial reform; police reform; counternarcotics; and disarmament, demobilization, and reintegration of former fighters, all led by separate coalition nations. The United States began efforts to build a new Afghan National Army (ANA). The initial

---

11 The division ready brigade was a light infantry brigade combat team that was worldwide deployable within 18 hours and capable of deploying by parachute assault. The author was part of the division ready brigade during that time period (2001-2002) as a cavalry troop commander. Degen and Reardon, Modern War in an Ancient Land: The United States Army in Afghanistan, 2001–2014, Volume 1, 236; Michael D. Fitzgerald, Notes from discussion at USCENTCOM headquarters, interview by Edmund J. Degen and Francis J.H. Park, March 18, 2015, OEF Study Group unclassified files, U.S. Army Center of Military History.

At the End of a 6,000-Mile Screwdriver   117

trainers for the ANA in 2002 were U.S. Special Forces, trained in advising.¹² Conventional force trainers from a light infantry brigade replaced the special forces, a mission eventually known as Task Force PHOENIX, whose remit also included defense institution building to support the creation of a professionalized military.¹³ After the initial failures of police reform under the Bonn process, the United States started a parallel program to train Afghan police.¹⁴

The Afghan army and police that were the ANDSF would not have been able to operate alone in any event, and securing the Afghan countryside itself was a task for the coalition itself while the ANDSF was being built. American policy direction eschewed nation-building, but expanding outside of the major bases in Kabul, Bagram, and Kandahar required establishing bases in new locations.

As ISAF expanded to the north, west, south, and east starting in 2003, it relied on troop-contributing nations to provide provincial reconstruction teams (PRTs) to assist in the stabilization effort supporting civilian governance and development. Those PRTs required military forces for sustainment and protection, and their presence expanded into regional command (RC) headquarters by 2006. Other than the United States, the troop commitments for every nation did not exceed that of a brigade task force.¹⁵ The more extensive scope of U.S. forces entailed an augmented infantry division headquarters to direct the activities of three brigade combat teams, special operations forces, and associated enablers, as well as serve as the U.S. national support element (NSE), which was responsible for the logistical sustainment of all U.S. forces in Afghanistan.

The presence of multiple troop-contributing nations in what looked to be a potentially lengthy mission brought with it the complications of multinational logistics. A long-running principle of multinational warfare is that logistics remains a national responsibility, and Afghanistan was no exception. Every country deploying troops to Afghanistan also deployed an NSE or made arrangements with other countries for sustainment.

---

12 Degen and Reardon, Modern War in an Ancient Land: The United States Army in Afghanistan, 2001–2014, Volume 1, 175, 227; Malkasian, The American War in Afghanistan, 126.
13 Reconstructing the Afghan National Defense and Security Forces: Lessons from the U.S. Experience in Afghanistan. (Arlington, Va.: Special Inspector General for Afghanistan Reconstruction., 2017), 15–22, https://apps.dtic.mil/sti/citations/AD1139856.
14 Reconstructing the Afghan National Defense and Security Forces, 26–28, 31.
15 Sten Rynning, NATO in Afghanistan: The Liberal Disconnect. (Palo Alto, CA: Stanford University Press, 2012), 47–52.

However, it was clear that any commitment of non-U.S. NATO forces would be contingent on U.S. logistical and combat support. In fact, some countries predicated their participation in ISAF on the presence of an American OEF corps headquarters in the country that could provide access to enablers such as close air support.[16]

Force composition in a multinational coalition is quite different than for a single national force. The countries that sent brigades to fill the RC headquarters generally also sent NSEs to sustain those units—except those NSEs counted against their country's force cap. Consequently, a multinational force always presented less actual combat capability than a similarly sized national contingent. That effect was compounded if a troop-contributing nation also had to guard its own base and did not share force protection with another country's forces.

In the initial years after ISAF expansion, this fragmentation of combat power became painfully apparent as the British and Canadians in RC(South) established bases in Helmand and Kandahar provinces, but they did not have enough combat power to consolidate gains made in combat because they did not have enough troops. The British sent a brigade with about two rifle battalions and a reconnaissance regiment to Helmand.[17] The Canadians in Kandahar had a brigade-size force with a rifle battalion's worth of combat power that routinely went 'outside the wire.' The United States initially had a division minus in RC(East), which expanded to over 30,000 by 2008.

Logistical flexibility within the Afghanistan theater of operations required sizeable capabilities in intratheater lift, enough combat forces to secure the bases and lines of communication (LOC), and the funding to sustain that force posture at expeditionary distances. The comparative lack of strategic mobility and the exorbitant costs of establishing and sustaining a logistical base, meant that most troop-contributing nations could establish only one base, which stayed until the end of the mission.

The one exception was the United States, whose logistical assets and combat power dwarfed those of every other country. In some cases, American logistical support kept certain countries in the fight through acquisitions

---

16 Rebecca Johnson and Micah Zenko, "All Dressed Up and No Place to Go: Why NATO Should Be on the Front Lines in the War on Terror," *Parameters*, Vol. 32, No. 4 (2002), 52–53, https://doi.org/doi:10.55540/0031-1723.2123; Degen and Reardon, *Modern War in an Ancient Land: The United States Army in Afghanistan, 2001–2014, Volume 1*, 175–76.

17 British Army reconnaissance regiments, commanded by a lieutenant colonel, are battalion-size organizations, unlike an infantry regiment that is comprised of multiple rifle battalions.

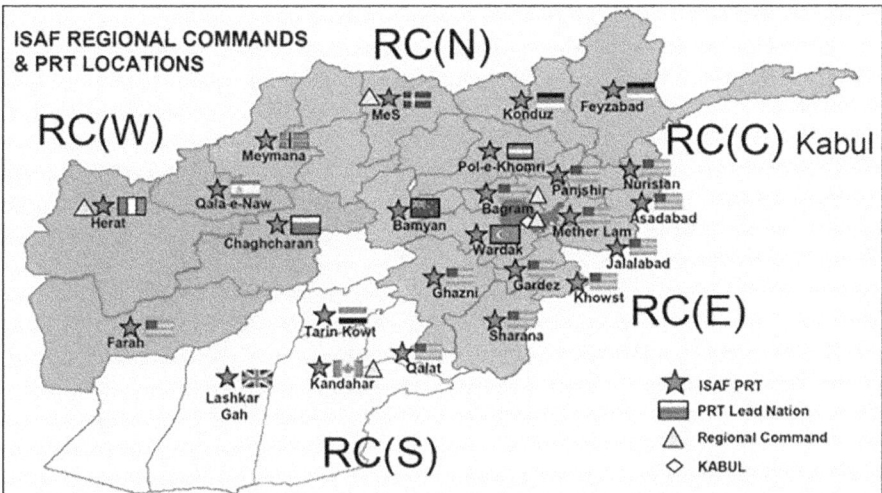

Figure 8.1: ISAF Regional Commands and Provincial Reconstruction Teams as of 5 December 2007[20]

and cross-servicing agreements, bilateral international instruments that allowed one country to provide cooperative logistical support to another.[18] In the meantime, there were numerous combat outposts and patrol bases that were established in the south and east to accommodate the increased footprint of American small units there, especially in Paktika, Gardez, Paktiya, Kunar, Khost, Nuristan, and Laghman provinces.[19] The density of presence was readily apparent in a 2007 NATO map showing the distribution of provincial reconstruction teams and major headquarters across Afghanistan.

Sustaining dispersed bases placed throughout a country of approximately 650,000 square kilometers placed a premium on LOCs, both external to Afghanistan or internal to it. Externally, the maturation of operational logistics after 2002 meant that there were effectively four GLOCs into Afghanistan: two in the north and two in the south. Some cargoes would

---

18 A particularly good example was in Zabul province, where a reinforced U.S. rifle company provided enablers that were critical to continued Romanian presence there. Wade C. Cleland, "Counterinsurgency Practices in An Economy-Of-Force Role: Zabul Province, Afghanistan 2006-2009" (master's thesis, Washington, DC, Georgetown University, 2010), 29.
19 A well-known example of one is Combat Outpost Restrepo in the Korengal Valley of Kunar Province, featured in the documentary films *Restrepo* and *Korengal*.
20 The author was the principal campaign planner for RC(East) from 2008-2009 and was responsible for revising campaign direction consistent with the change in ISAF's strategy during that time. Rynning, *NATO in Afghanistan*, 60.

only be transported via air due to problems with pilferage in Pakistan or political limitations on certain types of cargo on the northern GLOCs.

Internal to Afghanistan, the mountainous terrain in the country's center and along the southeast border with Pakistan meant that ground resupply of some outlying combat outposts was hazardous during the summer but virtually impossible during the winter, requiring aerial resupply of many bases. For the bases accessible by road, the American use of contracted trucking for common user logistics was critical in sustaining units at those outlying bases. At the same time, Bagram Airfield had become the central receiving point for strategic logistics coming from the U.S. theater sustainment command's assets in Kuwait.

2009-2014: Shift to Counterinsurgency Operations and Retrograde

The 2008 arrival of General David McKiernan entailed a major shift in the campaign strategy. McKiernan was the first ISAF commander to command OEF forces; thus, in military parlance, he was 'dual-hatted' as the NATO mission commander and the American mission commander. This arrangement mattered as the American mission was more combat-oriented. For example, the Americans in RC(East) conducted population-centric counterinsurgency operations since 2006. In 2007 the American approach started to accrete into ISAF's overall strategy. More significantly, under the new combined ISAF and OEF command arrangement, the overall effort focused on a comprehensive approach that viewed combat operations, security force assistance, and support to governance and development as mutually supporting efforts. General Stanley McChrystal succeeded McKiernan in September 2009, and he directed several major reforms, which in turn drove logistical requirements to handle the massive increase of American forces that became known as the Afghan Surge.

The roles of the different headquarters in Afghanistan evolved over time, which had significant effects on the management of the campaign. The ISAF headquarters, in addition to its role directing the campaign through political and military lines, was also the principal interlocutor to the Afghan civilian government and its military. The major increase in coalition forces in the Surge required the creation of an operational headquarters to handle the translation of campaign direction to tactical action, which resulted in the creation of the ISAF Joint Command (IJC). The second command and control change entailed the creation of NATO Training Mission-Afghanistan

(NTM-A) on top of the existing structure of the U.S. Combined Security Transition Command (CSTC-A), which oversaw the TF PHOENIX mission and defense institution building at the ministerial level.

The increasing level of violence in the theater since 2007 led McKiernan to request the deployment of an additional 30,000 troops to Afghanistan in 2008. American President Barack Obama formally announced that deployment in March 2009, and then he again increased the American presence in December 2009, making the total U.S. presence approximately 100,000 troops.[20] Operationally, the meteoric expansion of the American force structure in the Afghan Surge drove the changes to the logistical support for the campaign. The initial forces reinforced existing forces in RCs South and East, but U.S. brigade combat teams also started to operate in RC(North). The in-theater logistical structure also had to expand to the west and north.[21] The increased American strength and expansion of operational areas were challenging for planning and executing theater-level logistics; however, logistics remained a national responsibility, which offset some of these challenges. If the total increase in troops had come from across the coalition rather than just American, it would have been much more difficult to support logistically. The consequence was that the cost of that sustainment was borne solely by the United States, which was already providing ample logistical support to other countries' combat forces under acquisition and cross-servicing agreements.[22] As a result, many of those additional forces were not maneuver combat forces but enablers and sustainment forces to prevent culmination of the campaign.

Up to 2008, the Pakistan GLOC was the sole overland method of moving cargo (primarily non-lethal supplies and fuel) into Afghanistan. However, Islamabad prohibited American personnel from operating within Pakistan.[23] As a result, all cargoes offloaded from sealift in the port of Karachi moved on commercial carriers under nominal Pakistani control but through territory controlled by the Taliban and Haqqani Network. The resultant lack of in-transit visibility of cargoes transiting the Pakistan GLOC, combined with insurgent attacks and pilferage, resulted in the

---

21 Edmund J. Degen and Mark J. Reardon, eds., *Modern War in an Ancient Land: The United States Army in Afghanistan, 2001–2014, Volume 2.* (Washington: U.S. Army Center of Military History, 2021), 335.
22 MAJ Gregory A. Cannata, "Security Force Assistance with The Romanian–American Battle Group," *CALL Newsletter 11-06, Security Force Assistance Training by the JMRC*, December 2010, 17; Joint Chiefs of Staff, *JP 4-0 (2023)*, V–1.
23 MAJ Kerry Dennard, MAJ Christine A. Haffey, and MAJ Ray Ferguson, "45th Sustainment Brigade: Supply Distribution in Afghanistan," *Army Sustainment*, Vol. 42, No. 6 (2010), 13.

loss of almost 15 percent of cargoes. Trucking strikes were also a problem. Finally, the Pakistan GLOC was a potential single point of failure as it carried nearly 90 percent of non-lethal supplies going into Afghanistan.[24]

The unreliability of the Pakistan GLOC and the U.S. 2008-2011 Surge in Afghanistan led to the development of what became known as the Northern Distribution Network (NDN). The Defense Logistics Agency (DLA) and the U.S. Transportation Command (USTRANSCOM) began the work for the NDN in 2008 as part of planning for the Afghan Surge. That U.S. troop presence went from about 25,000 at the end of 2007 to 100,000 at the end of 2010.[25]

The NDN entailed three routes: NDN North, NDN South, and KKT (Kazakhstan, Kyrgyzstan, and Tajikistan). The NDN North began at the seaport of Riga, Latvia, then transited former Soviet rail lines in Russia, Kazakhstan, and Uzbekistan, where it was offloaded to trucks that entered Afghanistan at the border crossing point at Hairatan. The NDN South began at the seaport of Poti, Georgia, where cargo was trucked to Baku, Azerbaijan, and then ferried across the Caspian Sea to Aktau, Kazakhstan. From Aktau those cargos were trucked through Uzbekistan to Hairatan. The KKT line was an alternate route to the border crossing site at Sher Khan if the Uzbekistan route was blocked.[26] While Turkmenistan would have been an even more direct route, its isolationist government prevented all but humanitarian goods from transiting to Afghanistan.[27]

The NDN was an expensive tradeoff. Removing the Pakistan GLOC as a single point of failure and bringing secure transport of goods and the option of three additional routes into Afghanistan (not counting cargo moved by air) took 20 to 40 days of additional transit time at double

---

24 Craig Whitlock, "U.S. Turns to Other Routes to Supply Afghan War as Relations with Pakistan Fray," *Washington Post*, July 2, 2011, https://www.washingtonpost.com/world/national-security/us-turns-to-other-routes-to-supply-afghan-war-as-relations-with-pakistan-fray/2011/06/30/AGfflYvH_story.html; Tom Gjelten, "U.S. Now Relies on Alternate Afghan Supply Routes," National Public Radio, September 16, 2011, https://www.npr.org/2011/09/16/140510790/u-s-now-relies-on-alternate-afghan-supply-routes.

25 Gjelten, "U.S. Now Relies on Alternate Afghan Supply Routes"; Associated Press, "A Timeline of US Troops in Afghanistan since 2001," AP News, July 6, 2016, https://apnews.com/united-states-government-fe3ec7e126e44c728978ce9f4b5ebabd.

26 Andrew C. Kuchins and Thomas M. Sanderson, "The Northern Distribution Network and the Modern Silk Road: Planning for Afghanistan's Future." (Center for Strategic and International Studies, 2009), 8–11.

27 CPT Andrew P. Betson, "Nothing Is Simple in Afghanistan: The Principles of Sustainment and Logistics in Alexander's Shadow," *Military Review*, Vol. 92, No. 4 (2012): 56; Paul Stronski, "Turkmenistan at Twenty-Five: The High Price of Authoritarianism," Carnegie Endowment for International Peace, January 30, 2017, https://carnegieendowment.org/2017/01/30/turkmenistan-at-twenty-five-high-price-of-authoritarianism-pub-67839.

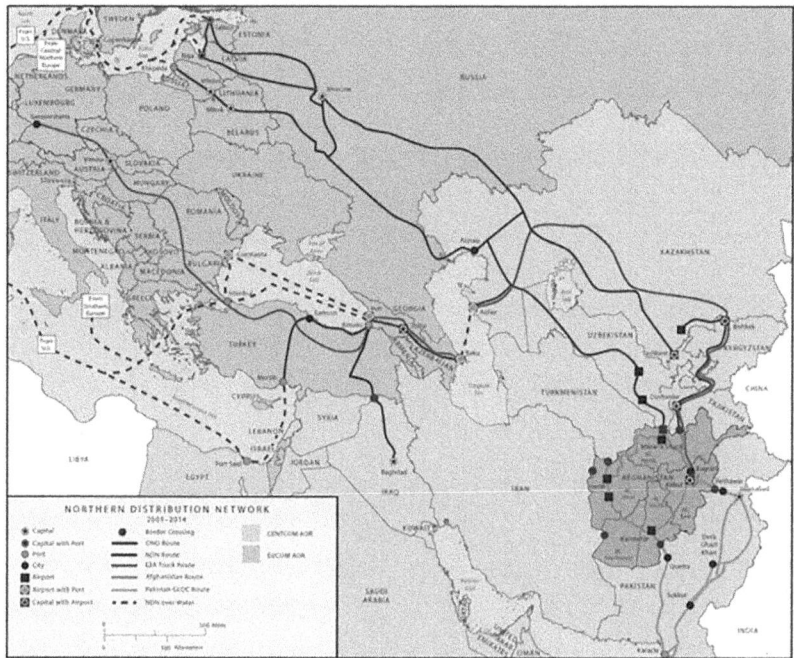

Figure 8.2: Northern Distribution Network Routes 2008-2014.[28]

the cost. Furthermore, it was an American solution to logistics, not a coalition solution. Consequently, while logistics was nominally a national responsibility, multiple ISAF countries depended on American national and theater logistics, primarily through "lift and sustain" agreements.[29] The complexity of the theater logistics infrastructure across two combatant commands came to require central direction from USTRANSCOM, which underscored the necessity for a single distribution manager to handle force sustainment entering the theater of operations. What also made it remarkable was how quickly American logistical throughput was able to adapt in support of the Surge and its influx of personnel and organizations. The scope of the NDN was immense, spanning multiple seaports and airports in Europe and Asia to sustain the theater.

However, the Afghan Surge was short-lived. Obama's December 2009 announcement telegraphed both the 18-month duration of the Surge and the end of the OEF mission at the end of 2014. By July 2011, as the last of the

---

28 Degen and Reardon, *Modern War in an Ancient Land: The United States Army in Afghanistan, 2001–2014, Volume 2*, 382–85.
29 Degen and Reardon, *Modern War in an Ancient Land: The United States Army in Afghanistan, 2001–2014, Volume 2*, 383.

Surge forces were still arriving, the troop withdrawals that became known as the "Surge Recovery" started, a process that would continue until the end of 2014. At the same time, the coalition began work to transition lead security responsibility to the ANDSF.[30]

The strategy to wind down OEF required several mutually supporting operational approaches. First were combat operations to protect the coalition force itself. The second was security force assistance and defense institution building to train, advise, and equip the ANDSF and its institutions. The third was the retrograde of forces and materiel out of Afghanistan—all required careful coordination given the size of the force and the number of moving parts involved.

The first two operational approaches required much closer coordination than had been the case in the past. With the impending time limit, coalition forces could no longer pay lip service to putting the ANDSF in the lead. That dynamic created a quandary where putting the less capable Afghans in the vanguard of combat operations entailed more tactical risk, but failure to do so meant the strategy would fail. The expansion of the U.S. presence to the north and west provided some of that capacity, while other countries' national caveats precluded such operations.

Any transfers of materiel from the coalition to the ANDSF would be tailored to what the Afghans could actually maintain, which ran counter to the rampant and long-standing culture of hoarding among many Afghan soldiers. What was required was weaning the Afghans off direct logistical support from the coalition. Additionally, the train, advise, and assist mission required institutional strategy skills that were not resident in tactical formations. For those reasons, the balance of the coalition's force structure would shift away from maneuver combat forces to security force assistance teams that provided enablers such as fires, medical evacuation, and intelligence support to the ANDSF.[31]

The third operational approach, while not directly related to the ANDSF, was still daunting. A decade of deploying forces to Afghanistan had generated a sizeable 'iron mountain' of materiel that had accumulated as units rotated in and out of theater. Adding to the complexity was that units had brought some of their own equipment from home station in

---

30 Degen and Reardon, *Modern War in an Ancient Land: The United States Army in Afghanistan, 2001–2014*, Volume 2, 377–79.
31 Degen and Reardon, *Modern War in an Ancient Land: The United States Army in Afghanistan, 2001–2014*, Volume 2, 395; "Report on Progress Toward Security and Stability in Afghanistan" (Department of Defense, July 2013), 13–15.

addition to receiving theater-provided equipment (TPE) that remained after the unit rotated out. While some of this materiel was in larger bases such as Bagram and Kandahar, much of it was dispersed throughout the country in less accessible locations. Units in the midst of combat operations faced the challenge of trying to account for materiel that had been in theater long before the unit arrived.[32]

Without a clear sense of asset visibility, the U.S. published an order to identify what TPE was present and whether it was serviceable. The initial estimate that followed was that moving 26,000 vehicles and 95,000 containers of equipment, valued at $25 billion, was $10 billion in shipping costs alone. One of the challenges of retrograding the theater was the scope of the withdrawal. Although the subsequent NATO mission (called RESOLUTE SUPPORT) to the ISAF and OEF missions was predicted to have a smaller presence after 2014, diplomatic approval from Kabul for the train, advise, and assist mission was not guaranteed. As a result, logisticians had to plan for a worst-case drawdown to 'zero.' As it turned out, that was the initial case for the retrograde, assuming that slowing the drawdown would be easier than accelerating it.[33]

Moreover, retrograding the theater had to occur on the same LOCs still being used to sustain the theater. Also, withdrawals that went through the Pakistan GLOC were subject to the same denial of access that had led to the creation of the NDN in the first place. After a border dispute with U.S. special operations forces in 2011, Pakistan closed the border from November 2011 to July 2012. Had the NDN routes not been available, the only recourse would have been to use airlift to move materiel out of Afghanistan. The Pakistani border closure resulted in impounding 10,000 containers, including food supplies discarded as wastage.[34]

The other challenge to retrograding was that logistics was still a national responsibility. Some countries, such as Germany, flew their cargo out of airfields they controlled, while others had to compete for limited space on shared airfields or on the NDN routes, which overlapped NATO agreements for ISAF. Some countries in that route had made agreements

---

32 Degen and Reardon, *Modern War in an Ancient Land: The United States Army in Afghanistan, 2001–2014, Volume 2*, 377.
33 The Bilateral Security Agreement and NATO Status of Forces Agreement that allowed coalition forces to remain in Afghanistan after 2014 would not be ratified until August 2014. Degen and Reardon, *Modern War in an Ancient Land: The United States Army in Afghanistan, 2001–2014, Volume 2*, 386–87.
34 Heidi Reisinger, "Not Only 'Containerspotting' - NATO's Redeployment from Landlocked Afghanistan," Research Paper. (Rome: NATO Defense College, October 2013), 7.

with NATO, which provided an umbrella for countries that did not negotiate their own bilateral agreements.[35]

By virtue of the size of its troop commitment (almost ten times that of the next largest contributor), the United States faced the most significant logistical issues. Division of labor for the disposition of U.S. property was divided across several organizations. The U.S. headquarters in Afghanistan had created a dedicated unit to handle the receipt of excess equipment to relieve tactical units from having to do so. At the same time, the U.S. Army Materiel Command identified equipment that was to be redistributed to other units in Afghanistan. Demilitarization and destruction of materiel slated for disposal rested with the U.S. Defense Logistics Agency (DLA), while USTRANSCOM was responsible for shipping the remainder to the United States. The delegation of authority for DLA to destroy serviceable equipment that was uneconomical to return to the United States significantly reduced the requirement to ship those cargoes.[36]

Unexpected events required logisticians to remain flexible. One example occurred in late 2013 when protesters from Imran Khan's Tehreek e-Insaf political party blocked the eastern route of the Pakistan GLOC after a member of the Pakistani Taliban was killed in a missile strike. In response, logisticians moved cargo on the Ring Road to Kandahar to the western route, while simultaneously staging cargoes on outbound cargo aircraft.

Over time, processes and modes of transport were established in the theater logistics infrastructure to accelerate the retrograde. One way to do so was to create a regular 'channel' flight that moved cargoes out of the theater of operations but still within U.S. Central Command's area of responsibility, where its theater sustainment command could handle subsequent disposition. Such a method required continued access to a theater logistics hub for materials entering or leaving Afghanistan, which the United States had in Kuwait. However expensive, the multimodal use of airlift to move materiel to Kuwait provided access to the two seaports and reduced the burden on the NDN or Pakistan GLOC.[37] By the end of 2014, the theater was set to transition to Operation FREEDOM'S SENTINEL and the NATO Resolute Support mission that followed ISAF.

---

35 Reisinger, "Not Only 'Containerspotting' - NATO's Redeployment from Landlocked Afghanistan," 7–9.
36 Degen and Reardon, *Modern War in an Ancient Land: The United States Army in Afghanistan, 2001–2014, Volume 2*, 387, 395, 397.
37 Degen and Reardon, *Modern War in an Ancient Land: The United States Army in Afghanistan, 2001–2014, Volume 2*, 419–23.

Insights from Theater Logistics in Afghanistan, 2001-2014

While an evaluation of the coalition's strategy is beyond the scope of this analysis, it drove certain decisions that had downstream effects on the logistics of the campaign. Despite the failures of strategy that resulted in the withdrawal of coalition forces from Afghanistan and the rapid disintegration of the ANDSF and the Kabul government in 2021, there are still valuable observations of logistics to be made from the campaign itself.

First, strategy, implemented through campaigning, determines what logistics is asked to do. That balancing of ends and means is a key aspect of the feasibility of a given strategy. While logistics may not be the dominant factor in determining the objectives of a campaign, it does bound what is within the realm of the possible. As the United States found out late in 2001 while attempting to set an austere theater, the ability to support the force was a limiting factor on the composition of the force that could be sent in the first place.

Improving the logistical capacity and capability in the theater was a continuous task but doing so required tradeoffs between combat forces and the sustainment forces, including national support elements. However, those improvements to set the theater were a necessary prerequisite to the introduction of the Afghan Surge forces in 2009. Much as before, deploying combat forces required a significant investment in enablers and sustainment forces. That investment, combined with U.S. capabilities for national strategic, theater strategic, and operational lift and distribution, was why the United States was the only troop-contributing nation whose forces could operate routinely in more than one regional command.

The limitation of other countries' bases to a single regional command reflected the inviolate reality that logistics was a national responsibility. Every country other than the United States effectively grew roots upon arrival to their base due to limitations in national-level capabilities of lift and distribution. The United States could support specific countries under acquisition and cross-servicing agreements, but those were purely bilateral instruments and not technically a form of coalition common logistics. However, U.S. support through the lift and sustain agreements offset some of those coalition shortfalls, especially during the retrograde of the theater.

Logistics and force flow were in no small part physics problems, as they were also governed by the iron hand of national policy. The relationship between the two is hard to overstate, especially in a coalition. Much of this challenge rested in LOCs and the form of the forces available for the campaign. The other place where that relationship became manifest was when making decisions, such as the retrograde out of theater, which were operationally difficult or tactically expensive but strategically necessary.

The confluence of the two appears in the form of the theater strategic and operational LOCs. The ability of the United States to draw upon its logistical hubs established in Kuwait and at the various air bases on the Arabian Peninsula was vital for setting the theater of operations. As the NDN became a more important route for cargo coming in and out of the theater, the European LOCs became vitally important. However, they were still subject to national interests, and the most direct route into Afghanistan was unavailable due to Turkmenistan's unwillingness to allow most cargo to transit. At the same time, Pakistan used closures of its GLOC as leverage on the coalition and the United States more directly, which the coalition could offset, albeit at significantly greater cost. In that case, the theater strategic infrastructure became vital to the United States as it retrograded materiel on the channel flight to Kuwait, which allowed staging those cargoes for sealift from Kuwait back to the United States.

Within the theater of operations, operational and tactical logistics were heavily reliant on American-provided common-user logistics. The preponderance of forces being American made sustainment easier for the coalition at large, since most of that sustainment could be handled within national means. In this case, national economies of scale made logistics feasible in ways that would not have been possible for a more balanced coalition force. Had the task organization been more balanced across troop-contributing nations, it would have been more difficult, if not impossible, to sustain dispersed units throughout the theater of operations, partly because of the necessity to open forward logistics bases, which underscored the realities of sustainment as a national responsibility, away from the U.S. theater logistics hubs at Bagram and Kandahar.

The last lesson to take from the role of logistics in Afghanistan is the importance of careful planning and execution. The confluence of competing national interests and the complexities of national strategic, theater strategic, and operational logistics required careful coordination of resources. The surge recovery and the retrograde really brought that need to the forefront. As difficult as it might be to open a theater of operations, it was just as challenging to close one. In the case of the former, an austere theater required discipline in what forces were being deployed and how they were sustained so that the campaign would not culminate. Closing the theater of operations and retrograding the force required similar levels of discipline to prevent culmination of the force on the way out or fratricide among stakeholders competing for the same lift assets and the same LOCs going out.

The war in Afghanistan from 2001 to 2014 was a unique conflict in many ways; however, like any other conflict, the iron hands of policy, strategy, and geography are non-negotiables, especially when fighting at expeditionary distance. Those challenges are amplified significantly when opening an austere theater of operations or having to close it. Having a coalition member supporting other members through lift and sustain agreements provides a backstop for the coalition at large, something that would be much more difficult in a more symmetric coalition. Many of the challenges that the United States and its coalition partners faced in Afghanistan will hardly unique, and operational reach and its considerations will remain an issue in future conflicts.

9

LOGISTICS, OPERATIONAL WARFARE, AND THE WAR IN UKRAINE

Jim Greer

As the Western powers concluded almost two decades of counterinsurgency operations in Iraq, Afghanistan, and other locations, they shifted their focus towards potential large scale combat operations.[1] In 2024, large-scale combat operations (LSCO) are no longer potential conflicts, but actual conflicts. Two years into the Russian Special Military Operation in Ukraine, almost a year into the war in Gaza, and with the ever-increasing potential of a Chinese attempt to take Taiwan by force, LSCO is a reality. By definition, large scale combat operations include large forces. They also almost certainly include combinations of primary opponent states, allies, partners, or proxies. And, most importantly for this chapter, large scale combat operations are inherently shaped by logistics. This chapter provides insights into the relationship between logistics and operational warfare—the conduct of campaigns and major operations—using the ongoing war in Ukraine as evidence.

In the late fall and early winter of 2021, the United States provided the Ukrainian and Western powers clear evidence that the Russian Federation intended to conduct a large-scale invasion of Ukraine.[2] While war in Ukraine had been ongoing between the Ukrainians and the Russians since 2014, in the summer and fall of 2021 Russian forces began massing across the entire boundary between Ukraine and Belarus and

---

1 King, James. Large-Scale Combat Operations: How the U.S. Army Can Get Its Groove Back. *Modern War Institute*. June 19, 2018. Downloaded at: https://mwi.westpoint.edu/large-scale-combat-operations-army-can-get-groove-back/.
2 Huminski, Joshua. Russia, Ukraine, and the Future Use of Strategic Intelligence. *Prism* (10)3. September 7, 2023. National Defense University Press. 9-25.

the Ukrainian territories in the Donbas the Russians had occupied since 2014. On February 24, 2022, the Russians initiated a large-scale invasion of Ukraine they termed a Special Military Operation (SMO). While most Western analysts assumed an all-out Russian *coup de main* would result in their defeating Ukraine in a matter of days, the opposite ensued. While the fighting has been incredibly intense, lethal, and resource consuming, more than two years later the Ukrainian Armed Forces continue to prevent Russia from accomplishing the objectives of their SMO. Throughout the course of this war, the strategy, operations, and tactics have been shaped by the logistics of Ukraine, supported by a broad coalition of more than 50 countries, and Russia, backed by Iran, North Korea, and China. The logistics efforts of both combatants, as well as their supporting nations, has shaped the strategy, operations and tactics throughout the conflict.

The focus of this chapter, operational warfare, is situated between strategy and tactics. At the strategic level, governments decide their policies, develop strategies, strengthen the will of the population and support of military action, and provide resources to enable military forces to conduct their operations in support of strategy. At the operational level, military forces, in conjunction with all the other elements of government and national power, engage in operational warfare to conduct campaigns and major operations designed to accomplish the strategic objectives of their state. Operational warfare consists of two components: operational art and operational science. The science of operations determines what can be done, while the art of operations suggests how it may be done. Campaigns or major operations are executed at the tactical level through the combination of battles, engagements, and other military activities. There is a synergistic relationship between the goals, decisions, and actions at the strategic, operational, and tactical levels.

It is important to understand that effective operational warfare combines both art and science. The science of operations, and especially the science of logistics, both enables and constrains the operational artist, that is, the operational-level commander. It signals to the operational artist what can and cannot be done, and for how long and how far military force can be applied. The operational artist designs campaigns and major operations that integrate logistics with various combinations of military forces and envisions how logistics will affect the operation over time.

Several of the elements of operation design are driven primarily by logistics. One of those is the idea of bases, from which military force is projected into and across a theater of operations.³ The second is the idea of operational reach.⁴ Operational reach is determined by how far logistics can support a campaign and for how long. And the third is the idea of culmination, which is the place and time that the logistics capacity of a force can no longer support offensive or defensive operations, and the force will be required to transition to some other operation.⁵ The fourth element of operation art that has a specific logistics focus is lines of operations, or the logistics grammar, lines of communications.⁶ These are the lines along which military forces and power are projected and sustained. Examples of lines of communication include road, rail, river, airports and air routes and seaports and sea lines of communications. The last element of operational design driven primarily by logistics is phasing.⁷ Campaigns and major operations are usually characterized by a number of phases in which primary activities are distinctly different. In the war in Ukraine logistics has been the primary, although not the only, driver of the transition from one phase to another. While each of these five elements of operational design is illustrated by the war in Ukraine, the general flow of the analysis and implications is structured by phase.

The Run-Up to LSCO

In the case of Russia's SMO in Ukraine, President Putin declared that Ukraine was not a state but was rather an integral part of Russia. The Russian strategy upon initiating the SMO was to undermine the Ukrainian government and military through intelligence and information operations and then conduct a *coup de main* by attacking with overwhelming force along several axes surrounding Ukraine to seize Kyiv and the other major cities, decapitate the government, and then annex Ukraine into Russia.⁸ That effort began on February 24, 2022, with a multi-domain operation that included air and missile strategic and operational attacks, an airmobile insertion to secure the Hozemel airport close to the Ukrainian capital, and

---

3   The Joint Staff. Joint Publication 5-0, *Joint Planning*. Washington, D.C.: U.S. Government. December 1, 2020. IV-8. Downloaded at: https://irp.fas.org/doddir/dod/jp5_0.pdf.
4   Ibid. IV-34.
5   Ibid. IV-28.
6   Ibid. IV-29.
7   Ibid. IV-37.
8   Watling, J., Danylyuk, O. and Reynolds, N. *Preliminary Lessons from Russia's Unconventional Operations During the RussoUkrainian War, February 2022–February 2023*. Special Report. Royal United Services Institute. 29 March 2023. 4-12.

major armor and mechanized thrusts along the axes of Belarus to Kyiv; Russia to Sumi, Chernihiv, and Kharkiv; and out of Crimea to Kherson; while at the same time attacking out of their positions they had occupied in the Donbas since 2014.

In contrast, the Ukrainian policy was one of distancing themselves from Russia and instead becoming a part of the European Union (EU) and, eventually, from a security standpoint, NATO. Their strategy was simply one of defense. Faced with the existential threat that Russia represented, the Ukrainian government and military forces simply wanted to defeat the initial Russian assault and retain their sovereignty. Their initial operational objectives across the more than 1000-kilometer front line were to blunt the Russian attacks and hold onto as much Ukrainian territory as they could. Once the Ukrainians had survived the initial onslaught, they were able to establish more proactive strategic objectives. Today, their stated strategic objective is to defeat the Russian SMO and restore the political borders that existed prior to 2014.[9]

During the months preceding the Russian offensive, the United States made it clear that no American troops would deploy to Ukraine to participate in its defense. Instead, support from the United States, other NATO nations, and other partners would be primarily through logistics and training. As the Russians massed forces of nearly 400,000 personnel on Ukraine's borders, the United States, NATO allies, and coalition partners began to increase their provision of weapons and munitions to Ukraine.[10] Thus, when the Russian attack started, initial logistics support to Ukraine was already underway. In the weeks before the invasion, Ukraine's allies and partners had donated significant amounts of munitions and other material and then transported them from strategic and operational distances into Ukraine.[11] In a relatively short period of time, the capabilities and capacity of the Ukrainian armed forces greatly improved through logistical support. The initial decision to mobilize their population enabled the Ukrainian Armed Forces to equip existing and new units and organizations with state-of-the-art weapons and equipment.

---

9 Jenson, B. and Hoffman, E. *Victory in Ukraine Starts with Addressing Five Strategic Problems.* Washington, D.C.: Center for Strategic and International Studies. May 15, 2024. https://www.csis.org/analysis/victory-ukraine-starts-addressing-five-strategic-problems.
10 U.S. Sent 30 Anti-Tank Missile Systems to Ukraine in October, Pentagon Says. Radio Free Europe. December 11, 2021. https://www.rferl.org/a/ukraine-30-anti-tank-javelin-systems/31604802.html.
11 Abramson, J. et al. *Arms Transfers to Ukraine*. Forum on the Arms Trade. Downloaded 6 July 2024 at https://www.forumarmstrade.org/ukrainearms.html.

The Initial Russian Offensive

When the Russian attack came on February 24, the Ukrainian units initially defended largely with old Soviet and Russian weapons and equipment, as they had been a member of the Soviet Union throughout the Cold War. During those first few days, Ukrainian air defense, artillery, tanks, infantry fighting vehicles, and individual weapons were largely Russian-made, and the ammunition they fired was from leftover stock from the USSR. Two areas where there was a significant difference were the provision of modern Western anti-tank guided missiles (ATGM)[12] and short-range air defense (MANPAD). As the regular Ukrainian Armed Forces brigades slowed and blunted the Russian attacks along the various axes, other Ukrainians, either newly equipped regular units or territorial forces units, could make good use of the ATGM and MANPAD provided by the West. Small groups of Ukrainians began roving along the length of and between the Russian columns, employing ATGM to ambush and destroy Russian tanks, infantry fighting vehicles, and other systems, including Russian supply convoys.

At the same time, individual Ukrainian soldiers, small units, and even police were able to employ Western-provided MANPAD to ambush and attack Russian fixed-wing aircraft and helicopters at lower levels. These MANPAD attacks drove Russian aircraft higher in the sky to avoid the short-range missiles, which in turn made those aircraft vulnerable to the Ukrainian Armed Forces Russian-made medium and high air defense systems.

The key here is that the myriad ATGM and MANPAD small engagements scaled up to create operational effects. That overall effect on the ground was to severely weaken the attacks out of Russia toward Chernihiv, Sumy, and Kharkiv and out of Belarus towards Kyiv. This approach contributed to the eventual defeat of those axes and the major Russian withdrawal after the first three months of the SMO.[13] At the same time, the effect in the air from western-provided MANPAD was to deny airspace to Russian air forces, which in turn preserved Ukrainian forces, capabilities, and freedom of action. The provision of logistics in the form

---

12 Sanger, D., Schmitt, E., Cooper, H., Barnes, J. and Vogel, K. Arming Ukraine: 17,000 Anti-Tank Weapons in 6 Days and a Clandestine Cybercorps. *The New York Times*. March 6, 2022. https://www.nytimes.com/2022/03/06/us/politics/us-ukraine-weapons.html.
13 Jones, S. *Russia's Ill-Fated Invasion of Ukraine: Lessons in Modern Warfare*. Washington, D.C.: Center for Strategic and International Studies, June 1, 2022. https://www.csis.org/analysis/russias-ill-fated-invasion-ukraine-lessons-modern-warfare.

of weapons and munitions contributed directly to success in defeating major objectives of the initial phase of the Russian SMO. In other words, logistics enabled multiple tactical engagements that, taken together, had an operational effect.

In terms of support, the Russian military forces operated from four major bases. To the north of Kyiv was Belarus, the base from which Russian forces attacked south to try and seize the capital of Ukraine. In the northeast was Belgorod, from which logistical support was projected along the Kharkiv, Sumi, and Chernihiv axis. In the southeast was Rostov-on-Don, from which military logistical support was projected into the previously occupied Donbas and the axis along the coast of the Sea of Azov to conduct the siege of Mariupol eventually. The fourth major logistical base was that of the Crimea. Having occupied Crimea in 2014, the Russian military had built up a significantly large logistical base from which they projected force north to eventually seize Kherson, as well as projecting force east along the coast of the Sea of Azov to contribute to the siege of Mariupol.

Basing is the reason that less than five weeks after the Russian SMO started the Ukrainians' very first cross-border attack into Russia was a helicopter attack to Belgrade.[14] The Ukrainians understood that they needed to disrupt the logistical bases of the Russian armed forces to slow down and weaken their advance during the early months of the war. Weakening the Russian Armed Forces ability to project combat power from Belgorad would in turn affect their operational reach.

Operational reach is the distance that a force and its combat power can be projected during a campaign or major operation. Logistics is the primary determinant of operational reach, while the range of weapons systems (such as long-range fires and aircraft) can extend operational reach to a certain extent. That said, while long-range fires and aircraft can extend the distance that combat power can be applied, they cannot by themselves secure terrain. Thus, operational reach in campaign terms is dependent on the ability to move ground forces and support those forces in combat operations.

In the case of the Russian SMO, the operational reach of the attacking force was limited by the range of their supporting logistics transportation. The Russian armed forces depend heavily on logistics from existing bases

---

14 Kwon, J., Angelova, M., Hodge, N. and Pavlova, U. *Russia accuses Ukraine of helicopter strikes on fuel depot in Russian territory.* CNN, April 1, 2022. https://www.cnn.com/2022/04/01/europe/russia-ukraine-belgorod-fire-intl/index.html.

while moving vast quantities of supplies, vehicles, and munitions forward by rail. At the beginning of the offensive, the Russian transportation trucks were limited in quantity and range. The available trucks could only supply Russian attacking forces about 40 miles from the last base or railhead. Since the Russians had no railroads in Ukrainian territory, that meant supplying from the bases in Belarus, Russia and Crimea. The Russian operational reach then was about 40 miles, which contributed to the attacks along the various axes culminating when they did.

The initial Russian attack along the line of operation from Belarus due south to Kyiv used a single line of communication. Consequently, when the Ukrainians were able to block the Russian advance along that route, the Russian forces backed up, creating the famous 40-kilometer-long convoy. With their logistics fuel and ammunition stuck behind the combat forces on the road, the Russian attack could not continue.[15] The Ukrainians continually attacked and attritted that column, and the Russians eventually abandoned this column. After only six weeks the Russian main effort had culminated. That is, the Russian attack toward Kyiv had reached the point where it no longer had sufficient combat power and sustainment to maintain the offensive and was forced to transition and withdraw forces from the attack toward the Ukrainian capital.

In a similar fashion, the Russian attacks out of the Belgorod base toward Sumy, Chernihiv, and Kharkiv failed for much the same reasons. The major roads serving as supply lines through northeastern Ukraine were limited on each axis, creating logistics bottlenecks.[16] Those supply lines were then subjected to attacks by roving Ukrainian elements armed with Western-supplied ATGMs, resulting in the same culmination forced on the Kyiv axis. Along all three axes the Russian forces transitioned to repositioning their over-extended forces, giving up virtually all of the Kharkiv and Sumy Oblasts they had seized in early February.

---

15 Ti, R. and Kinsey, C. Lessons from the Russo-Ukrainian Conflict: The Primacy of Logistics Over Strategy. *Defence Studies*, 23:3, 381-398, DOI: 10.1080/14702436.2023.2238613. https://doi.org/10.1080/14702436.2023.2238613.
16 Martin, B., Barnett, S. and McCarthy, D. *Russian Logistics and Sustainment Failures in the Ukraine Conflict*. Research Report. Washington, D.C.: RAND. January 2023.8. https://www.rand.org/content/dam/rand/pubs/research_reports/RRA2000/RRA2033-1/RAND_RRA2033-1.pdf.

The Second Russian Offensive

After the failure of the original offensive, the Russians repositioned, giving up the terrain they had lost because of an inability to project logistics in the north. Almost immediately, they renewed their assault in the northern Donbas, seeking to secure the entirety of the Luhansk Oblast. Having adapted from their failed first offensive, rather than attempting rapid armored drives deep into Ukraine, they reverted to a more methodical, attrition-based approach. The Russian Army has always been artillery-based, and throughout its history, it has employed artillery in mass to destroy defenses and enemy units.[17] In the summer of 2022, the Russian army adopted the tactics that had worked in the past. Over several months, the Russian forces were able to take some terrain, particularly the city and logistical hub of Severodonetsk.[18] They did this simply by using artillery to pulverize the city one block at a time. After a portion of the city was leveled, their infantry would move in and occupy. This tactic was repeated block by block until the Ukrainian defenders finally withdrew from the city.

The Russians were able to employ these tactics because their logistics provided them with an overwhelming advantage in artillery. At this point in the war, the Ukrainians were beginning to run out of Soviet-era artillery ammunition and had not yet started to receive large quantities of either NATO artillery or 155mm ammunition. The Russian army was firing up to 60,000 shells and rockets per day, while the Ukrainians could not respond in kind.[19] However, that advantage was short-lived. In the summer of 2022, Ukraine received its first U.S.-supplied High Mobility Artillery Rocket Systems (HIMARS). HIMARS, with its precision and range, was able to attack the Russian artillery dumps and logistics nodes.[20] In a period of just a few weeks, HIMARS significantly reduced the Russian artillery advantage in the Donbas. The Russians were forced to adapt by moving their ammunition far from the front and only bringing it forward in small amounts. Here again, logistics had a significant impact on the campaign.

---

17 Bellamy, C. *Red God of War: Soviet Artillery and Rocket Forces*. London, UK: Brassey's. 1986.
18 Zabrodskyi, M., Watling, J., Danylyuk, O. and Reynolds, N. *Preliminary Lessons in Conventional Warfighting from Russia's Invasion of Ukraine: February–July 2022*. Special Report. Royal United Services Institute. November 30, 2022. 41-42.
19 Schogol, J. Russia is hammering Ukraine with up to 60,000 artillery shells and rockets every day. *Task and Purpose*. June 13, 2022. https://taskandpurpose.com/news/russia-artillery-rocket-strikes-east-ukraine/.
20 Douro, M. *MLRS and the Totality of the Battlefield*. London, UK: Royal United Services Institute. February 20, 2023. https://www.rusi.org/explore-our-research/publications/commentary/mlrs-and-totality-battlefield.

This phase of the war concluded with the Russians having secured some terrain but having once again been blunted by the Ukrainians, mainly because of logistics.[21]

The First Ukrainian Counter-Offensive

In the fall of 2022, the Ukrainians had stabilized the defenses all along their line of contact with Russian forces. At the same time, combat systems, munitions, and other supplies began to flow in large numbers from the nations that donated to Ukraine. As a result, the Ukrainian Armed Forces (UAF) was able to conduct offensives in both the Kharkiv and Kherson regions. After signaling an attack in Kherson for several weeks, the Kharkiv offensive took advantage of surprise to regain 6000 square kilometers of territory they had lost earlier that year.[22] In part, this attack was successful because the UAF was able to attack logistics depots and headquarters to set conditions for their offensive.[23] The logistical flow from the U.S., other NATO nations, and other coalition partners was starting to have an operational effect. By September of 2022, the U.S. alone had provided 1,500 MANPAD missiles to Ukraine.[24] Air defense provided by coalition members, moving along lines of communication starting in Poland and other border countries and moving by rail and road to the front lines, enabled the UAF to drive the Russian air forces back behind their own lines and enable an offensive largely free from interference from the air.[25]

These lines of communication from Europe to the front lines in Ukraine travel along a supply chain divided into three phases. The first phase was strategic as it brought the equipment, munitions, or supplies from the point of origin to the border with Ukraine. The second phase was operational in nature, transited the border, and moved across Ukraine to a distribution point near a particular front, in the case of the first Ukrainian

---

21 Martin, B., Barnett, S. and McCarthy, D. 10-11.
22 The Economist. *A stunning counter-offensive by Ukraine's armed forces*. September 15, 2022. https://www.economist.com/europe/2022/09/15/a-stunning-counter-offensive-by-ukraines-armed-forces.
23 Glantz, M. *How Ukraine's Counteroffensives Managed to Break the War's Stalemate*. Washington, D.C.: United States Institute of Peace. September 19, 2022. https://www.usip.org/publications/2022/09/how-ukraines-counteroffensives-managed-break-wars-stalemate.
24 U.S. Department of Defense. Fact Sheet on US Security Assistance to Ukraine. September 8, 2022.
25 Bronk, J., Reynolds, N. and Watling, J. *The Russian Air War and Ukrainian Requirements for Air Defence*. Special Report. London, UK: Royal United Services Institute. November 7, 2022. 14. https://static.rusi.org/SR-Russian-Air-War-Ukraine-web-final.pdf.

counter-offensive either Kharkiv or Kherson. The last phase was tactical and consisted of retail distribution (in this case of STINGER Air Defense Missiles) down to the unit or Soldier.[26]

The Kharkiv offensive also illustrated that basing is not limited to strategic and operational bases for power projection. In conflict, bases serve a critical role in enabling a campaign to proceed. Throughout this war, there have been repeated battles for various cities. Cities are important for a number of reasons, including the information value to the political leadership of both sides, the fact that they are population centers where people live, and their operational impact. That operational impact is often because cities serve as logistical bases to project tactical and operational combat power and forces. Several examples of that from this war are the cities of Lyman and Izium. Both were important in the initial Russian invasion because they were logistical hubs and a center where multiple lines of communication crossed. For the Russians to attack and secure the northern Donbas, they had to take Lyman and Izium, which they did early in their initial offensive. And it is for that very reason that when the Ukrainians were in the position to counterattack during the fall of 2022, their primary objectives were the towns of Lyman[27] and Izium.[28] Thus, a key outcome of their counter-offensive was retaking Lyman, denying its use to the Russians as an operational base, and providing their own base for defending the northern Donbas.

In the southern portion of the Ukrainian first counter-offensive, Kherson is also an example of a base with tactical and operational implications. In the initial offensive in 2022, one of the Russian operational objectives was to attack out of Crimea, through Kherson and Mykolaiv, and eventually to Odessa. Because of effective Ukrainian defense, however, the Russian offensive stopped between Kherson and Mykolaiv, but the Russians did secure a significant portion of Kherson Oblast, north of the Dnipro River. In the fall of 2022, in conjunction with the offensive in Kharkiv, the UAF attacked south toward the Dnipro River and retook

---

26 Castillo, V. How weapons get to Ukraine and what's needed to protect vulnerable supply chains. *The Conversation*. March 16, 2022. https://theconversation.com/how-weapons-get-to-ukraine-and-whats-needed-to-protect-vulnerable-supply-chains-179285.
27 CNBC. *Ukraine claims full control of key logistics hub, eyes further gains*. October 2, 2022. https://www.cnbc.com/2022/10/02/ukraine-claims-full-control-of-key-logistics-hub-of-lyman-eyes-further-gains.html.
28 Reuters. *Russia gives up key northeast towns as Ukrainian forces advance*. September 10, 2022. https://www.reuters.com/world/europe/ukraine-troops-raise-flag-over-railway-hub-advance-threatens-turn-into-rout-2022-09-10/.

the city of Kherson. This eliminated Kherson as a base of operations for the Russian armed forces.

With major Russian forces defending north of the Dnipro River, the Ukrainians used long-range fires to isolate the battlefield. Given that all Russian lines of operations, and therefore logistics support, must come across the river, the UAF attacked the bridges themselves. At the same time, the Ukrainian attack itself was along the right bank of the river gradually cutting off the Russians from their supply lines.[29] Despite Russian resistance, in-depth positions, and thousands of mines, the UAF finally reached the city. By then, the Russian Army was evacuating the city. To ensure the Ukrainian offensive culminated at the river, the main bridge from Kherson south, the Antonivka Bridge, was blown up and collapsed on November 11, 2022, just after the last Russian troops escaped south. After the Russian retreat, the Ukrainians were able to collect significant amounts of weapons, ammunition, and supplies that they would use in later operations.[30] Thus, the second phase of the Russian SMO, the first Ukrainian Counter-Offensive, consisting of major offensive operations in the Kharkiv and Kherson regions, was shaped at the operational level by logistics factors of basing, lines of operations, operational reach, and culmination.

The Second Ukrainian Counter-Offensive

Following the successful Ukrainian operations of the fall of 2022, the UAF prepared to conduct further offensive operations while simultaneously defending the city of Bakhmut. UAF received modern Abrams, Leopard, and Challenger tanks; Bradley and other infantry fighting vehicles; and self-propelled artillery to form new brigades that would be employed in the 2023 counter-offensive.[31] But, along with the flood of new NATO equipment came the logistical challenges of maintaining, supplying, and

---

29 Sciutto, J. and Lister, T. *Ukrainian forces aim to retake Kherson by year's end as gains made in south, US and Ukrainian officials say.* CNN. September 7, 2022. https://edition.cnn.com/2022/09/07/politics/ukraine-russia-war-kherson/index.html.
30 Cotovio, V., Kiley, S., Rudden, P. and Konovalova, O. *Inside the Battle for Kherson.* CNN. November 21, 2022. https://edition.cnn.com/2022/11/21/europe/ukraine-kherson-battle-intl/index.html.
31 Axe, D. *The Ukrainian Army Could Form Three New Heavy Brigades with All These Tanks and Fighting Vehicles It's Getting.* Forbes. January 17, 2023. https://www.forbes.com/sites/davidaxe/2023/01/17/the-ukrainian-army-could-form-three-new-heavy-brigades-with-all-these-tanks-and-fighting-vehicles-its-getting/.

transporting a myriad of different model tanks and armored vehicles. At the same time, the Ukrainians had to maintain and supply their own, largely former Soviet, equipment along with significant numbers of captured Russian systems.[32]

When the Ukrainian counter-offensive commenced, rather than concentrating on a single axis, the UAF attacked Bakhmut in the east and toward Tokmak in the south. The attack toward Bakhmut was a supporting effort to regain some of the terrain lost to the Wagner Group in Bakhmut and divert Russian attention away from the main effort toward Tokmak. However, this forced the UAF to operate along multiple extended lines of operations that extended from the NATO countries to the west all the way to the front lines. This weakened both UAF attacks and contributed to the early culmination of the main effort toward Tokmak.[33]

The Third Russian Offensive

In the fall of 2023 after the Ukrainian counter-offensive had culminated, the Russian Army returned to the offensive. Through the summer of 2024, two major Russian efforts emerged. The first is a continued grinding advance in the Donbas, characterized by Russian advantages in artillery and manpower. Few, if any, new logistics lessons can be learned there. The second effort was different, the reopening of an offensive along the axis from Belgorod toward Kharkiv. Here the Russian Army took advantage of the Ukrainian border and the United States and NATO's restrictions against firing NATO-provided munitions across the border. In effect, this created a self-imposed sanctuary within which the Russian Army could position its logistics nodes and operate protected lines of communication.[34]

However, after the initial Russian assault, the U.S. and other NATO nations changed their policy, allowing limited use of provided munitions

---

32 Aleksic, M., Cvrk, S. and Bozivic, D. Analysis of Land Army Maintenance Techniques in the War in Ukraine. *Military Review*. May-June 2023, 33-45. https://www.armyupress.army.mil/Journals/Military-Review/English-Edition-Archives/May-June-2023/Land-Army-Maintenance-Techniques/.
33 Watling, J., Danyluk, O. and Reynolds, N. *Preliminary Lessons from Ukraine's Offensive Operations, 2022–23*. London, UK: Royal United Services Institute. July 2024.
34 Barros, G. Putin's Safe Space: Defeating Russia's Kharkiv Operation Requires Eliminating Russia's Sanctuary. *Kyiv Post*. May 14, 2024. https://www.kyivpost.com/post/32609.

to attack Russian logistics and forces within Russia herself.[35] The effect was to degrade Russian logistics to the point Russian Soldiers complained of a lack of food. The UAF was able to blunt the offensive and conduct local counterattacks to restore a portion of their boundary and limit the Russian gains.[36] Given the limited nature of the Russian attack toward Kharkiv, the elements of operational reach and culmination have not been a factor.

The Changing Character of Logistics

The first two plus years of the Russian SMO have confirmed the logic of accounting for basing, lines of operation, operational reach, and culmination, and those elements impact phasing when planning for, executing, and assessing operational warfare. At the same time, the changing character of war demonstrated by the Russo-Ukraine War suggests that the character of logistics is also changing at the operational level. The science of new approaches to warfighting and logistics affects the application of the operational art in this conflict, the ongoing war in Gaza,[37] and future conflicts.

One of the important factors of the changing character of the conflict by the second year was the growing impact of small uncrewed aerial systems (sUAS) or drones. In the first year of the Russian SMO, sUAS played a crucial tactical role. They served as reconnaissance both to find and target the enemy. They also served as attack systems, usually by dropping small grenades or explosives. But, due to the limited availability of such drones, during 2022, they had only tactical effects. In the second year of the war that changed. In the initial operations of the UAF counter-offensive, the Ukrainians were losing ten thousand drones a month.[38] That meant that they were using at least ten thousand drones a month. As 2023

---

35 Kallub, S. As US-Supplied Weapons Show Impact Inside Russia, Ukrainian Soldiers Hope for Deeper Strikes. *Military.Com*. June 23, 2024. https://www.military.com/daily-news/2024/06/23/us-supplied-weapons-show-impact-inside-russia-ukrainian-soldiers-hope-deeper-strikes.html.

36 O'Grady, S., Khudov, K. and Morgunov, S. Ukrainian attacks on supply lines slowed Russians in Kharkiv, intercepts show, *The Washington Post*. https://www.msn.com/en-us/news/world/ukrainian-attacks-on-supply-lines-slowed-russians-in-kharkiv-intercepts-show/ar-BB1pAgdy?ocid=BingNewsSerp.

37 Frantzman, S. *New tools of war: How the IDF uses drones to fight Hamas in Gaza*. Foundation for Defense of Democracies. January 27, 2024. https://www.fdd.org/analysis/op_eds/2024/01/27/new-tools-of-war-how-the-idf-uses-drones-to-fight-hamas-in-gaza/.

38 Watling, J. and Reynolds, N. Meatgrinder: *Russian Tactics in the Second Year of Its Invasion of Ukraine*. Special Report. London, UK: Royal United Services Institute. May 19, 2023. https://static.rusi.org/403-SR-Russian-Tactics-web-final.pdf.

progressed, First Person View (FPV) and bomber drones emerged as UASs became a significant component of the combat power of both combatants. And, like ATGMs and STINGERS, many drone attacks scaled up to create operational effects. The operational effect in this case was to make offensive operations that much more challenging to conduct, creating a stalemate, or, at best, minimalizing gains by either side throughout 2023 and the first half of 2024.[39]

By 2024 the Ukrainians and Russians both stated they would field a million drones a year. Those drones must be moved from where they are made to the front lines. If they are actually built at the front lines, as is the case with many of the Ukrainian drones, then the components must be moved from wherever they are purchased to the front lines. Drone batteries must also be bought and moved to the locations that require them. The drones that are not destroyed must be maintained or repaired. The support systems for drones, the control stations, launch platforms, and communications relays must all be moved, positioned, and maintained. Thus, achieving operational effects as they have in late 2023 and 2024 requires a whole logistics, infrastructure, and support system unique to the drones.[40] And, this is overlaid on top of all the existing logistical systems on both sides.

Moreover, the UAF illustrates the hybrid nature of drone warfare and drone logistics. Ukrainian drones are manufactured by their government, by defense contractors, by small businesses, by non-profit organizations, by individual units, and by individuals themselves.[41] This complex approach to developing and deploying small unmanned aircraft systems (sUAS) requires a similarly complex approach to logistics. Traditional military logistics processes and formations are insufficient for the variety, rapidity, and adaptation required to employ the myriad drone types and operations.[42]

---

39 Cropsey, S. *Drone Warfare in Ukraine: Historical Context and Implications for the Future.* Strategika. The Hoover Institution. March 14, 2024. https://www.hoover.org/research/drone-warfare-ukraine-historical-context-and-implications-future.
40 Kunertova, D. Drones Have Boots: Learning from Russia's War in Ukraine. *Contemporary Security Policy.* 44 (4), 576-591. https://www.tandfonline.com/doi/epdf/10.1080/13523260.2023.2262792?needAccess=true.
41 Temple-Raston, D. and Powers, S. Exclusive: Inside Ukraine's secret drone factories. *The World.* October 23, 2023 https://theworld.org/stories/2023/10/23/exclusive-inside-ukraine-s-secret-drone-factories.
42 Pettyjohn, S. Evolution Not Revolution: Drone Warfare in Russia's 2022 Invasion of Ukraine. Center for New American Security. February 2024 https://s3.us-east-1.amazonaws.com/files.cnas.org/documents/CNAS-Report-Defense-Ukraine-Drones-Final.pdf.

A second area in which the character of logistics is changing is that of contested logistics on the battlefield. Heretofore, contested logistics usually referred to that in the home country or during strategic deployment. For example, contesting logistics within the continental United States or while in transit across one of the oceans. However, on the battlefield of Ukraine, logistics are more contested at tactical and operational distances from the Forward Line of Troops. Historically, logistics were interdicted deep by airpower or long-range fires, and even then, the inability to target logistics nodes or lines of communication limited effectiveness accurately.

In the Russian SMO that changed. Ubiquitous, effective sensors from space to air to ground to the electromagnetic spectrum (EMS) make hiding extremely difficult. The proliferation of sUAS and long-range loiter munitions have made attacking logistics more effective than in the past.[43] Add to that the effectiveness of long-range precision strikes against logistics (remember the effects of HIMARS in significantly degrading Russian artillery logistics in the summer of 2022). All this requires that logistics change to be more survivable through techniques such as dispersion, deception, and reorganization.[44]

Conclusion

Operational warfare—the art and science of designing and executing campaigns and major operations—depends entirely on logistics. Over the last two plus years, the Russian SMO has demonstrated that statement is as true today as it was in World War II. Logistics on both sides has driven what is possible and what is not and shaped the character and operations of each phase of the Russo-Ukraine War. While the traditional logistics-centric elements of operational design (bases, lines of communications, operational reach, and culmination) have had significant impact, new considerations have emerged. The first of those is many, discrete, small logistics actions *scaling up* to create operational effects, as the provision of ATGM and MANPAD did in the first phase of the Russian SMO. The second is the emerging reality of *hybrid logistics*, integrating traditional

---

43 Fox, A. Contested Logistics: A Primer. Landpower Essay 24-1. Association of the United States Army. February 2024. https://www.ausa.org/sites/default/files/publications/LPE-24-1-Contested-Logistics-A-Primer.pdf.
44 Hertlein, R. Army Logistics Survivability Against Multidomain Threats. U.S. Army. February 23, 2023. https://www.army.mil/article/263363/commentary_army_logistics_survivability_against_multidomain_threats.

logistics with public-private partnerships, non-profits, and even crowd-sourced efforts. The third of these considerations is that of tactical and operational *contested logistics* having an immediate impact on battles, major operations, and phases of a campaign. A continued detailed study of logistics in the war in Ukraine that produces lessons and changes behaviors is necessary if friendly nations and forces are to succeed in operational warfare in the future.

PART III

FUTURE
BRAVE NEW LOGISTICAL WORLD

10

TOWARD A THEORY OF THE SUPPLY CHAIN ENVIRONMENT

Matt Evers

The supply chain environment continues evolving as an increasingly complex system shaping how states, societies, and militaries interact. The nature of the environment and likely threats have introduced new terms like contested logistics[1] and the contested logistics environment[2] into the vernacular of security and defense circles. However, new definitions are only part of changing strategy, force design, capability development, and operational concepts necessary for success in this new environment—accompanying theory is needed. New theory is required as the terms logistics, contested logistics, and contested logistics environment lack consensus of definition across communities of practice, and the larger supply chains and their systematic activities in which the military logistics function occurs do not always consider the larger supply chains.[3] It is also redundant to label a function as contested if the environment in which that function is inherently contested as a complex adaptive system and subject to the fog and friction of war.

---

1 Amos C. Fox, "Contested Logistics: A Primer," *Landpower Essay* 24-1. (Arlington, VA: Association of the United States Army, February 2024), 1, https://www.ausa.org/publications/contested-logistics-primer.
2 Jon Michael King, "Contested Logistics Environment Defined," *Army Sustainment* Winter 2024 (February 2024), https://www.army.mil/article/272922/contested_logistics_environment_defined.
3 For example, the Association for Supply Chain Management, a global leader in supply chain thought leadership and training defines logistics as: 1) In a supply chain management context, it is the subset of supply chain management that controls the forward and reverse movement, handling, and storage of goods between origin and distribution points. 2) In an industrial context, the art and science of obtaining, producing, and distributing material and product in the proper place and in proper quantities. 3) In a military sense (where it has greater usage), its meaning can also include the movement of personnel. APICS, Inc., *ASCM Supply Chain Dictionary Seventeenth Edition* ed. Paul H. Pittman and J. Brian Atwater (Chicago, IL: ASCM, 2022), 109, https://www.ascm.org/learning-development/certifications-credentials/dictionary/.

Nonetheless, contested logistics is prolific and has energized a buzz of activity across organizations in and out of the U.S. Department of Defense (DoD); however, there are several issues.[4] First, there is no common language rooted in the supply chain management discipline among civil servants, the military, and industry. The community also lacks a shared understanding of the contested environment between the public and private sectors. Third, there is a common acceptance of the problems to be managed. Certainly, executives across government, the civilian leadership of the DoD, think tanks, and professionals in the commercial sector are collectively understanding the complexity, national security risks, and China's malign influence on supply chains and are aggressively moving out in promulgating strategy and making change as fast as governmental politics can allow.[5] However, military leaders will be limited in their ability to contribute to strategy without a foundational understanding of supply chain theory expressed through updated doctrine.

The need is clear. A theory of the supply chain environment requires a common language for military logistics, changing the military's

---

4 Megan Gully, "Army focuses on contested logistics – a threat to enemy," news release, April 3, 2023, https://www.army.mil/article/265428/army_focuses_on_contested_logistics_a_threat_to_enemy.; Army Futures Command, *VERTEX Contested Logistics February 13-14, 2024, Information Sheet* (Austin, TX: Army Futures Command, 2024). https://vertex.aal.army/contested-logistics/; and Tom Karako, "Project Convergence: An Experiment for Multidomain Operations" (panel transcript, February 16, 2022), Center for Strategic & International Studies, https://www.csis.org/analysis/project-convergence-experiment-multidomain-operations.

5 Joseph Clark, "DOD Releases First Defense Industrial Strategy," news release, January 12, 2024, https://www.defense.gov/News/News-Stories/Article/Article/3644527/dod-releases-first-defense-industrial-strategy/; U.S. Department of Defense, *Securing Defense-Critical Supply Chains* (Washington, DC: Department of Defense, February 2022), 7-8, https://media.defense.gov/2022/Feb/24/2002944158/-1/-1/1/DOD-EO-14017-REPORT-SECURING-DEFENSE-CRITICAL-SUPPLY-CHAINS.PDF;John G. Ferrari and Mark Rosenblatt, *Preparing Supply Chains for a Coming War*, American Enterprise Institute (Washington, DC: AEI, February 2024), 4, https://www.aei.org/wp-content/uploads/2024/02/Preparing-Supply-Chains-for-a-Coming-War.pdf?x85095; John Perez, CPSM, CPSD, "Your Supply Chain Isn't Ready for the Next War," *Inside Supply Management Magazine*, September 12, 2023, https://www.ismworld.org/supply-management-news-and-reports/news-publications/inside-supply-management-magazine/2023-september/your-supply-chain-isnt-ready-for-the-next-war/; Shyam Sankar, Greg Little, and Madeline Zimmerman, "From Last Supper to First Breakfast: The Defense Tech Ecosystem," *Palantir (Medium)*, September 8, 2023, https://blog.palantir.com/from-last-supper-to-first-breakfast-cb971128b0bf; *Industry Perspectives on Defense Innovation and Deterrence: Hearing before the House Armed Services Subcommittee on Cyber, Information Technologies, and Innovation*, 118th Cong., (September 20, 2023) (statement of James D. Taiclet, Chairman, President, and CEO, Lockheed Martin Corporation); and The White House, *Executive Order on America's Supply Chains: A Year of Action and Progress*, (Washington, DC, February 2022), 3, https://www.whitehouse.gov/wp-content/uploads/2022/02/Capstone-Report-Biden.pdf; Graham Allison and Philip Zelikow, *Essence of Decision: Explaining the Cuban Missile Crisis*, 2nd ed. (Addison Wesley Longman, 1999), 255-256.

dominant logic and refining its doctrinal grammar. Otherwise, military leaders will continue to focus too narrowly on tactical distribution and materiel problems that address the symptoms but not the causes within the larger contested environment. A narrow approach would result in not recognizing institutional blind spots to areas of vulnerability and opportunity; furthermore, it would erode the military's readiness, credible deterrence, preparedness for protracted conflict, and ability to counter malign influence.

A new military logistics theory starts with reframing logistics from an enabling function focused on supply and transportation to the broader context of deeply interconnected supply chains as a maneuver space where military advantage is lost or gained, much like the emergence of the information environment and equivalent to the physical domains of warfare. With the emergence of these domains as a point of departure, contested logistics can be transformed into a unified and deeper understanding of the supply chain environment as a complex system and enlighten military leaders toward points of leverage in the environment for relative advantage in future conflict. With this understanding, the U.S. military can play a more strategic role in deterring adversaries and, if necessary, prevailing in conflict.

Background

States, societies, militaries, and industries recognized the emergence of the supply chain environment in the wake of unforeseen supply chain crises that have left the world reeling, such as the COVID-19 pandemic and the blockage of the Suez Canal by the *Ever Given* container ship.[6] Consequently, the supremacy of American logistics that had underwritten the U.S. military's lethality, deterrence options, and strategic flexibility is eroding, marked by strategic competition, rapid technological change,

---

6   Elisabeth Braw, "What the Ever Given Taught the World," *Foreign Policy* (November 10, 2021), https://foreignpolicy.com/2021/11/10/what-the-ever-given-taught-the-world/; and Council of Economic Advisors, The White House, *Issue Brief: Supply Chain Resilience*, (Washington, DC: November 30, 2023), https://www.whitehouse.gov/cea/written-materials/2023/11/30/issue-brief-supply-chain-resilience/.

and global supply chain risk.[7] The greater national security community is raising concern over the erosion of the competitive advantage of U.S. logistics. This challenge includes the previously unforeseen defense-critical supply chain dependencies, where third-tier suppliers of U.S. munitions and microelectronics are controlled by Chinese firms.[8]

U.S. power projection and logistics depend on access, basing, and overflight and civ-mil cooperation; however, subversive posturing by China through dual-use overseas port and facility investments enable China to delay U.S. forces and extend global influence.[9] The U.S. support to Ukraine to counter Russia's protracted war has depleted U.S. ammunition reserves and stressed domestic defense industrial capacity.[10] Others have noted the institutional underinvestment in combat service support platforms, logistics training, and the infrastructure to support weapon system sustainment at a rate that has diverged well off the pace of emerging technology and anti-access, area-denial threats.[11]

Threat-informed operational concepts by the military services demand intense and protracted operations over great distances in a more distributed and transregional fight, requiring substantial changes to military logistics. For example, the U.S. Marine Corps had to wholesale rewrite the logistics chapter of the second edition of the *Tentative Manual for*

---

7   Joint Chiefs of Staff, *Joint Logistics*. Joint Publication 4-0 (Washington, DC: Joint Chiefs of Staff, July 20, 2023), I-3 to I-5; Joint Readiness and Seapower and Projection Forces Subcommittee Hearing: Posture and Readiness of the Mobility Enterprise – TRANSCOM and MARAD, 118th Cong., (March 28, 2023) (opening written statement of General Jacqueline D. Van Ovost, Commander, US Transportation Command), 4-6, https://armedservices.house.gov/sites/republicans.armedservices.house.gov/files/General%20Van%20Ovost%20Witness%20Statement.pdf; and The White House, *National Security Strategy*, (Washington, DC., October 2022), 8-10, https://www.whitehouse.gov/wp-content/uploads/2022/10/Biden-Harris-Administrations-National-Security-Strategy-10.2022.pdf.
8   Jeffrey Jeb Nadaner and Tara Murphy Dougherty, *Numbers Matter: Defense Acquisition, U.S. Production Capacity, and Deterring China* (Washington, DC: Govini, 2024), 7, https://govini.com/research/numbers-matter-2024/.
9   Isaac B. Kardon and Wendy Leutert, "Pier Competitor: China's Power Position in Global Ports," *International Security* 46, no. 4 (Spring 2022), 43, https://doi.org/10.1162/isec_a_00433.
10 Seth G. Jones, *Empty Bins in a Wartime Environment: The Challenge to the U.S. Defense Industrial Base*, Center for Strategic & International Studies (Washington, DC: CSIS, January 23, 2023), 2, https://www.csis.org/analysis/empty-bins-wartime-environment-challenge-us-defense-industrial-base.
11 Chris Dougherty, *Buying Time: Logistics for a New American Way of War*, Center for a New American Security (Washington, DC: CNAS, April 2023), 1, https://www.cnas.org/publications/reports/buying-time.

Expeditionary Advance Base Operations.[12] The U.S. Navy continues to explore how to link its largest advanced naval base—the industrial conglomerates in the continental United States—with its most forward fleets to execute Distributed Maritime Operations.[13] The U.S. Air Force had to shift a mission generation paradigm built around the efficiency of main operating bases to more resilient, lightweight, and interconnected contingency locations and multi-capable Airmen to execute Agile Combat Employment.[14]

These new operational concepts have severe logistical constraints from the strategic to tactical level, requiring further analysis and wargaming to inform force design and capability development.[15] Recent analytical reports for each of these concepts recognize that logistics cannot be viewed in isolation as a supply and transportation problem of material but must intertwine related logistics activities, including matters of personnel, information, finance, and technology development.

The military's hyperfocus on tactical distribution and materiel problems stems three-fold from long-running issues of military thinking on logistics before the term contested logistics was trending. First, a lack of consensus on the definition and scope of logistics in preparing for and conducting war often focused only on matters of supply and transportation.[16] The interdependencies and incongruencies of military logistics at scale with government and commercial activity was another problem.[17] Third, an incomplete intellectual framework for military

---

12 Brian Kerg, "A Summary of Changes in the New EABO Manual," *Proceedings* 149/7/1445 (U.S. Naval Institute: July 2023), https://www.usni.org/magazines/proceedings/2023/july/summary-changes-new-eabo-manual.

13 Joslyn Fleming, Bradley Martin, Fabian Villalobos, and Emily Yoder, *Naval Logistics in Contested Environments: Examination of Stockpiles and Industrial Base Issues* (Santa Monica, CA: RAND Corporation, March 6, 2024), 43-45, https://www.rand.org/pubs/research_reports/RRA1921-1.html.

14 Patrick Mills, James A. Leftwich, John G. Drew, Daniel P. Felten, Josh Girardini, John P. Godges, Michael J. Lostumbo, Anu Narayanan, Kristin Van Abel, Jonathan W. Welburn, and Anna Jean Wirth, *Building Agile Combat Support Competencies to Enable Evolving Adaptive Basing Concepts* (Santa Monica, CA: RAND Corporation, April 16, 2020), xiii-xvii, https://www.rand.org/pubs/research_reports/RR4200.html.

15 Travis Reese, Curtis Hudson, and Matt Evers, *Contested Logistics Wargaming: A Report on the 8-10 August 2023 AMCL Symposium Wargaming Panel & Wargame Design Workshop*, Association of Marine Corps Logisticians (AMCL: Prepared September 30, 2023), 13, https://www.marinecorpslogistics.org/amcl-2023.

16 Kenneth Macksey, "A Revolution in Complexity: The Price of Neglect," in *For Want of a Nail: The Impact on War of Logistics and Communications* (United Kingdom: Brassey's, 1989), 5; and Milan Vego, "Operational Logistics," in *Joint Operational Warfare: Theory and Practice*, Reprint of 1st ed. (Newport, RI: U.S. Naval War College, 2009), VIII-75.

17 Henry E. Eccles, "Concepts of Logistics," in *Military Concepts and Philosophy* (New Brunswick, NJ: Rutgers University Press, 1965), 71-76.

commanders to judge and anticipate the future operational environment, especially with respect to technological change.[18] Given the complexity of these conditions, definitions, and conceptions of contested logistics and the contested logistics environment alone will always fall short without an accompanying theory. An introduction of the supply chain environment theory and a working definition of the supply chain environment is a start.

Logistics Theory Development

While definitions can provide common terms of reference to guide practice in a discipline, knowledge building and use are incomplete without theory.[19] The design challenge of defining supply chain and logistics is similar to the trouble in defining what war is, where the baseline theory is widely (and cross-culturally) accepted from the military theorist Carl von Clausewitz' assertions that war is a clash of wills and a continuation of policy with other means. In contrast, new definitions and additive theories are measured through the careful study of the changing character of war.[20]

The narrow understanding of military doctrine of logistics as a tactical function focused internally on supply and transportation is also a legacy of another 19th century military theorist, Baron Antoine-Henri Jomini. Jomini needed and 'invented' a term—logistics—to put meaning to the calculated movement of armies along geographic lines of operation to link the terrain-oriented objectives derived from military strategy to the execution of tactics upon those objectives and to sustain those operations.[21] However, a closer reading of Jomini and the *Art of War* suggests that Jomini may have understood supply chain management theory, the upstream activities of national production/supply, and the downstream activities of military consumption/demand, although there was no contemporary vocabulary for these notions.

---

18 Andrew F. Krepinevich, "Calvary to Computer: The Pattern of Military Revolutions," *The National Interest* Fall 1994, no. 37 (1994), 41-42, https://www.jstor.org/stable/42896863.
19 Carles L. Owen, "Design Research: Building the Knowledge Base," *Design Studies* 19, no. 1 (January 1998), 9-20, retrieved from https://id.iit.edu/wp-content/uploads/2015/03/Design-researching-building-the-knowledge-base-Owen_desstud97.pdf.
20 Hew Strachen, "The Elusive Meaning and Enduring Relevance of Clausewitz," in *The New Makers of Modern Strategy: From the Ancient World to the Digital Age*, ed. Hal Brands (Princeton, NJ: Princeton University Press, 2023), 133.
21 Antoine Baron de Jomini, *The Art of War*, trans. G. H. Mendell, USA and W. P. Craighill (Kingston, Ontario: Legacy Press, 2008), 46-47.

Jomini classified the supply chain activities of national defense as military institutions that included having a national capacity to develop (and steal if necessary) superior military technology, the acquisition and development of human capital, secret war reserve materiel, supply chain intelligence to determine a competitor's war reserve materiel, and financial capacity.[22] He asserted, "We must admit that a happy combination of wise military institutions, of patriotism, of well-regulated finances, of internal wealth and public credit, imparts to a nation the greatest strength and makes it best capable of sustaining a long war."[23] Today, Jomini's thinking falls short in his conception that these supply chain activities are internal functions to control when the contemporary reality of hyperconnected global supply chains characterizes these activities as part of a larger, complex environment that can only be influenced.

In 1917, Lieutenant Colonel George C. Thorpe, USMC, published *Pure Logistics*, a critical logistical treatise in which Thorpe argued that the interrelated activities of logistics could not be viewed in isolation, that they required national direction for the formulation of strategy and policy, and, most importantly, that the development of human capital—education— was a part of logistics.[24] Therefore, logistics meant more than optimizing the movement of men and materiel; maintaining and elevating the term helped unite interrelated disciplines into a comprehensive field. However, this early effort met little acclaim and fell to the wayside, a tragic and harmful output at this time.

After World War II, Rear Admiral Henry E. Eccles, the "grand old man of naval logistics,"[25] had a much greater success than Colonel Thorpe in starting a 'bow wave' of intellectual attention on logistics and war, organizational change in the U.S. Navy,[26] logistics professional military education, and the institutionalization of his logistics theories in U.S. military doctrine.[27] In 1965,

---

22 Ibid., 29-31.
23 Ibid., 29-32.
24 Stanley L. Falk, "Introduction," in *George C. Thorpe's Pure Logistics: The Science of War Preparation*, (Washington, DC: National Defense University Press, 1986), xxii-xxviii.
25 U.S. Naval War College Archives, "Henry E. Eccles papers," https://www.usnwcarchives.org/repositories/2/resources/95.
26 Peter C. Luebke, Timothy L. Francis, and Heather M. Haley, *Contested Logistics: Sustaining the Pacific War* (Washington, DC: Naval History and Heritage Command, 2023), 80-81, https://www.history.navy.mil/content/dam/nhhc/research/publications/publication-508-pdf/contested_logistics_508.pdf.
27 Scott A. Boorman, "Fundamentals of Strategy – The Legacy of Henry Eccles," *Naval War College Review* vol. 62, no. 2 (Spring 2009), 9-10, https://digital-commons.usnwc.edu/nwc-review/vol62/iss2/8.

he published *Military Concepts and Philosophy*, building off his previous works, and *Operational Naval Logistics* and *Logistics in the National Defense*—all three works based on his experience from World War II and the Korean War. He desired to give new meaning to military doctrine for future conflict, as being under the shadow of nuclear war and the permeating business mindset of the McNamara defense administration where "the military lost control of their own language; and the scholars, scientists, and publicists—who were without command experience and responsibility—moved in."[28]

Eccles recognized that military [logistics] theory required a more strategic, civil-military, and multi-dimensional scope, often starting from a simple philosophical perspective that wrestled with complexity, the tensions of political-military thinking, and the paradox within principles of war. Eccles expanded the meaning of logistics further as a 'bridge' between the national economy and military force and that the resource scarcity between the production side of the "economic system of the nation" and the consumption side of the "tactical concepts and environment of the combat forces" required a special kind of harmony to match requirements with production.[29] His analogy characterizes logistics as the link between strategy and tactics, like Jomini, but in a much grander sense.

Eccles also described logistics as a living contradiction, with his 'snowball' principle, where the scope and scale of logistics can grow so large that a "huge accumulation of slush obscures the hard core of essential combat support and the mass becomes unmanageable."[30] Eccles referred to the span of control and technical expertise for military logistics being so unwieldy that it would invite unwanted bureaucratic structure that would undermine combat effectiveness and, more strategically, deterrence. He exclaimed, "One of the chief weapons in the hands of those who advocate the increase of centralization and of civilian control of logistics has been the charge that military commanders are not competent to control their own logistics."[31]

What the 'bridge' and 'snowball' metaphors suggest is that the scope and scale of military logistics are so broad and an inextricable part of warfare that for military commanders to exercise good military judgment does not require a depth of technical knowledge of logistics functions, but

---

28 Henry E. Eccles, "Introduction" in *Military Concepts and Philosophy* (New Brunswick, NJ: Rutgers University Press, 1965), 8.
29 Ibid., 72.
30 Ibid., 83.
31 Ibid., 102.

a depth of understanding of the complexity of logistics as a more extensive system. Reframing logistics from an enabling function to supply chains as a unique aspect of the operational environment, like a physical domain of warfare, underscores the importance to military commanders just as knowledge of terrain. What is needed by military leaders is a broader and deeper understanding of military logistics and supply chain management theory to apply operational art for future conflict.

Supply Chain Theory Development

As logistics-related activities expanded in technical breadth and depth, business consultants introduced the term supply chain in the 1980s, expanding to academia as a unified field of study in the 1990s. In 2015, Carter et al. addressed the problem of advancing the supply chain management discipline with a seminal work that described what everyone had been discussing and trying to manage—the supply chain itself. Carter *et al.* proposed six foundational principles to describe the nature of supply chains, paraphrased as a network of friction and flow; a complex adaptive system comprising a clash of interests, disorder, fluidity, and risk; and bounded by a 'fuzzy horizon' of uncertainty and impeding terrain. Through this environment ran three fundamental flows—physical, informational, and financial.[32]

The baseline theory highlighted that logisticians well understood the physical flow and scope of the supply chain (supply and transportation), but the cross-section of the informational and financial flows and their associated supply chain activities were an under researched and under managed area, a 'known unknown,' let alone what else may be 'unknown unknowns' due to the supply chain's 'fuzzy horizon' and changing character. As described by Carter *et al.*, the nature of supply chains characterizes them as a unique and evolving environment, much like the information environment and not a function, which is supply chain management. Today's information systems and digital technologies have changed the character of militarized great power rivalries and ushered in

---

32 Carter, Craig R., Dale S. Rogers, and Thomas Y. Choi, "Toward the Theory of the Supply Chain." Journal of Supply Chain Management 51, no. 2 (April 2015), 89.

a fourth industrial revolution and next-generation digital supply chain.[33] However, the military has yet to adopt the widely accepted meaning of supply chains in its theory and doctrine, instead retaining Jomini's logistics lexicon focused on the movement of men and material.

The Supply Chain Operations and Reference (SCOR) model is an authoritative reference for supply chain management. This tool was introduced in 1996, overhauled in 2001 to account for the Internet, and revised in 2020 to account for emerging technologies like artificial intelligence.[34] The 14th version of the SCOR model published in 2022 integrates people as a focus area alongside supply chain performance, processes, and practices; frames supply chains as a nonlinear and constantly moving system; and expands stratagems for supply chain resilience, economy, and sustainability.[35] That said, supply chain management theory is about as young as joint military doctrine about cyber warfare, the information environment, and competition and should be carefully examined to advance the practice of military logistics and operational art.

Of note, the supply chain management theories that predominantly focus on the inventory of material is marginal compared to the majority of supply chain management theories that focus more strategically on competitive advantages, microeconomics, systems, and the intersection of the supply chain with the cognitive dimension, like marketing, sociology, psychology, decision making, and innovation.[36] The literature review above suggests that supply chain management theory can be applied to a much more comprehensive array of military operations across the competition continuum beyond the enabling function of logistics.

Furthermore, there are several popular planning constructs and evolving models from commercial supply chain management theory to inform strategic supply chain design around principles like agility,

---

33 Knut Alicke, Jürgen Rachor, and Andreas Seyfert, "Supply Chain 4.0 – the next-generation digital supply chain," *McKinsey & Company, Operations* (October 27, 2016), https://www.mckinsey.com/capabilities/operations/our-insights/supply-chain-40--the-next-generation-digital-supply-chain#/; and Michael J. Mazarr, *Understanding Competition: Great Power Rivalry in a Changing International Order — Concepts and Theories*. (Santa Monica, CA: RAND Corporation, 2022), 18, https://www.rand.org/pubs/perspectives/PEA1404-1.html.
34 Association for Supply Chain Management, *The ASCM SCOR Digital Standard Information Model* (ASCM, 2020), 2-3 and 30, https://scor.ascm.org/api/files/25?v=1714034767824.
35 Ibid.
36 C. Clifford Defee, Brent William, Wesley S. Randall, and Rodney Thomas, "An inventory of theory in logistics and SCM research," *The International Journal of Logistics Management* vol. 21, no. 3 (2010), 411, DOI 10.1108/09574091011089817.

resiliency, and efficiency; however, research has shown that senior military leaders who are responsible for defense supply chain strategy are wholly unfamiliar with the logic and grammar of supply chain management.[37] This lack of familiarity may reflect a gap in learning culture and human capital investment in military logisticians as junior leaders transition from a tactical to a strategic scope of responsibility with a myopic understanding of supply chain operations.[38] The difficulty for military leaders in understanding supply chains also lies in the nature of supply chains being a complex adaptive system, where "supply networks merge rather than result from purposeful design by a singular entity."[39] The scale, hyperconnectivity, and complexity of contemporary supply chains make them more than a function, but a unique environmental factor that cannot be entirely controlled. Theory is therefore crucial in understanding complex adaptive systems because these systems are rife with nonlinearities and emergent factors, where patterns may emerge that are counterintuitive and actions to influence the system can be unpredictable.[40] Therefore, military logistics and supply chain doctrine should be reframed from a linear enabling function to a unique supply chain environment.

A description of supply chains that highlights their current scope, scale, and evolving character provides a common term of reference for the emerging issues with supply chains and the proposed theory herein. There is wide divergence and a lack of consensus on the definitions of supply chain and supply chain management due to the nature of definitions being too broad or too narrow, their arbitrary standardization making it challenging to translate and contextualize across communities of practice, and the definition remaining elusive as the evolving complexity of supply chains are emergently recognized by practitioners.[41] But quite simply, supply chains are not just about the supply of material.

---

37 Thomas Ekström, Per Hilletofth, Per Skoglund, "Differentiation strategies for defence supply chain design," *Journal of Defense Analytics and Logistics* vol. 4, no. 2 (October 2020), 192, DOI 10.1108/JDAL-06-2020-0011.
38 Robert E. Overstreet, Joseph B. Skipper, Joseph R. Huscroft, Matt J. Cherry, and Andrew L. Cooper, "Multi-study analysis of learning culture, human capital, and operational performance in supply chain management: The moderating role of workforce level," *Journal of Defense Analytics and Logistics* vol. 3 no. 1 (March 2019), 52, DOI 10.1108/JDAL-11-2018-0017.
39 Thomas Y. Choi, Kevin J. Dooley, and Manus Rungtusanatham, "Supply Networks and Complex Adaptive Systems: Control Versus Emergence," *Journal of Supply Chain Management* vol 19, no. 3 (May 2001), 351, DOI 10.1016/S0272-6963(00)00068-1.
40 John H. Holland, *Hidden Order: How Adaptation Builds Complexity* (Helix Books, 1995), 5.
41 Steve LeMay, et al., "Supply chain management: the elusive concept and definition," *The International Journal of Logistics Management* 28, no. 4 (April 2017), 1445, DOI 10.1108/IJLM-10-2016-0232.

Physical goods require scarce productivity inputs like commodities, infrastructure, labor, machinery, and technology to transform and distribute products across points of production and consumption, forming commerce, trade, and economies.[42] Furthermore, supply chain products are not only physical things, but digital things like software, with cyber versions of transformation and distribution processes that introduce new threat vectors.[43] Supply chain products can also be cognitive things, like the talent and skills in people that require a support system for human capital acquisition, development, and retention.[44] People and organizations within the supply chain self-organize, interact, and orchestrate resources based on the flow of information (or lack thereof), from the discovery of emergent factors to signals about the market forces of supply and demand, to regulatory measures, to the fidelity of transactions across organizations and systems.[45] Finally, a financial system enables the exchange of products and services across the supply chain and the financial capacity to adapt productivity inputs or have purchasing power.[46]

Thus, supply chains have a much broader context than material and act as a network of physical, informational, financial, and human capital flows across nodes of agents and activities that seek to add value

---

[42] Wang Dongfang, Pablo Ponce, Zhang Yu, Katerine Ponce, and Muhammad Tanveer, "The future of industry 4.0 and the circular economy in Chinese supply chain: In the Era of postCOVID19 pandemic," *Operation Management Research* 15 (May 2022), 342-344, https://doi.org/10.1007/s12063-021-00220-0.

[43] Enduring Security Framework, Critical Infrastructure Partnership Advisory Council. *Securing the Software Supply Chain: Recommended Practices Guide for Developers* (August 2022), 1-2, https://www.cisa.gov/sites/default/files/publications/ESF_SECURING_THE_SOFTWARE_SUPPLY_CHAIN_DEVELOPERS.PDF; and Maoyang Wang, Peng Wu, and Qin Luo, "Construction of Software Supply Chain Threat Portrait Based on Chain Perspective," *Mathematics* 2023, 11, 4856 (December 2023), 9, https://doi.org/10.3390/math11234856.

[44] Association for Supply Chain Management, *ASCM Supply Chain Operations Reference Model: SCOR Digital Standard* (ASCM, 2022), xi, https://scor.ascm.org/api/files/24?v=1714034767824; and Robert E. Overstreet, Joseph B. Skipper, Joseph R. Huscroft, Matt J. Cherry, and Andrew L. Cooper, "Multi-study analysis of learning culture, human capital, and operational performance in supply chain management: The moderating role of workforce level," *Journal of Defense Analytics and Logistics* vol. 3 no. 1 (March 2019), 51-53, DOI 10.1108/JDAL-11-2018-0017.

[45] Mohita Gangwar Sharma, "Supply chain, geographical indicator and blockchain: provenance model for commodity," *International Journal of Productivity and Performance Management* vol. 72, no. 1 (2023), 92-95, DOI 10.1108/IJPPM-05-2021-0288.

[46] Chaorui Huang, Felix T.S. Chan, and S.H. Chung, "Recent contributions to supply chain finance: towards a theoretical and practical research agenda," *International Journal of Production Research* vol. 60, no. 2 (2022), 507-508, https://doi.org/10.1080/00207543.2021.1964706.

for customers and other stakeholders through products and services.[47] The combination of numerousness, connectivity, emergence, and human interests in the supply chain network make it more than a function, but a complex adaptive system that surrounds an individual—an environment. The distinction of environments as a complex adaptive system is critical as it implies a different logic to how individuals may assert control, achieve desired influence, and rebalance power.

Supply chains are also systems of interconnected processes. One of the leading professional bodies in the field, the Association for Supply Chain Management (ASCM), updated its reference model in 2022 and identified seven macro processes that makeup supply chains and operations.[48] The two demand-side macro processes are ordering products and services and fulfilling customer orders through various distribution activities.[49] The two supply-slide macro processes are sourcing products and services and transforming products and services through maintenance, development, and manufacturing.[50] The three integrating macro processes at the intersection of supply and demand are orchestrating the integration and enablement of supply chain strategies, planning across all functions, and the returning or reverse flow of goods, services, or components.[51]

ASCM's model is an evolution from the 1998 definitions and theory of supply chain and logistics from the Council of Logistics Management, at which point logistics became widely considered by the commercial sector as a subset of supply chains focused on what are currently the fulfill and return macro processes of the ASCM model, making the terms logistics and supply chain no longer synonymous.[52]

However, the latest 2023 U.S. joint doctrine inversely labels logistics

---

47 Craig R. Carter, Dale S. Rogers, and Thomas Y. Choi, "Toward the Theory of the Supply Chain," *Journal of Supply Chain Management* 51, no. 2 (April 2015), 89; and Douglas M. Lambert, Martha C. Cooper, and Janus D. Pagh, "Supply Chain Management: Implementation Issues and Research Opportunities," *The International Journal of Logistics Management* vol. 9, no. 2 (July 1998), 1, https://doi.org/10.1108/09574099810805807.
48 Association for Supply Chain Management, *ASCM Supply Chain Operations Reference Model: SCOR Digital Standard* (ASCM, 2022), iv-v, https://scor.ascm.org/api/files/24?v=1714034767824.
49 https://scor.ascm.org/processes/order; and https://scor.ascm.org/processes/fulfill.
50 https://scor.ascm.org/processes/source; and https://scor.ascm.org/processes/transform.
51 https://scor.ascm.org/processes/orchestrate%20supply%20chain; https://scor.ascm.org/processes/plan; and https://scor.ascm.org/processes/return.
52 Douglas M. Lambert, Martha C. Cooper, and Janus D. Pagh, "Supply Chain Management: Implementation Issues and Research Opportunities," *The International Journal of Logistics Management* vol. 9, no. 2 (July 1998), 3, https://doi.org/10.1108/09574099810805807.

and supply chain. U.S. Joint doctrine classifies supply chain management as a function focused primarily on material and subordinate to one of the core joint logistics functions, supply operations.[53] The language difference can confuse orchestrating and planning supply chain strategies and activities across the DoD, the interagency, commercial partners, and allies. Interestingly, the doctrinal publication does not include the term contested logistics at all, although significant updates speak to multi-domain operations, strategic competition, and technological change.[54] Joint logistics is also subordinate to an umbrella military term, the joint warfighting function, sustainment.[55]

U.S. Army doctrine also uses the term sustainment as an umbrella term, but its ontology of associated elements and functions aligns more closely with commercial supply chain management standards than U.S. Joint doctrine, particularly with the inclusion of financial management and personnel services.[56] Clarification of the doctrinal terms does matter, as the U.S. Army's interchangeable use of the terms logistics, combat service support, and sustainment with inconsistent definitions and changing hierarchies up until FM 3-0 was published in 2008 elicited political infighting and organizational inefficiency.[57] The new doctrinal definition of sustainment marked such a milestone in institutional learning and organizational learning that the U.S. Army changed the name and branding of its flagship periodical, *Army Logistician* to *Army Sustainment* at the publication's 40th anniversary in 2009.[58]

Applied Supply Chain Environment Theory

Doctrine expresses a military's dominant logic, worldview, and culture, influencing how the institution prepares for and conducts war. A working

---

[53] Joint Chiefs of Staff, *Joint Logistics*. Joint Publication 4-0 (Washington, DC: Joint Chiefs of Staff, July 20, 2023), II-3.
[54] Joint Chiefs of Staff, *Joint Logistics*. Joint Publication 4-0 (Washington, DC: Joint Chiefs of Staff, July 20, 2023), iii.
[55] Ibid., ix.
[56] U.S Army, *Sustainment*, ADP 4-0. (Washington, DC: Headquarters, Department of the Army, July 31, 2019), v-vi, https://armypubs.army.mil/ProductMaps/PubForm/Details.aspx?PUB_ID=1007565.
[57] Jeffrey C. Brlecic, "Logistics, CSS, Sustainment: Evolving Definitions of Support," *Army Sustainment* vol. 41, 5, (September-October 2009), 22, http://www.alu.army.mil/alog/2009/SepOct09/pdf/asust_septOct_09.pdf.
[58] Robert D. Paulus, "*Army Logistician* to Army Sustainment: Continuity and Change," *Army Sustainment* vol. 41, 5, (September-October 2009), 2, http://www.alu.army.mil/alog/2009/SepOct09/pdf/asust_septOct_09.pdf.

doctrinal definition for the supply chain environment is *the aggregate of material, informational, financial, and human capital factors that affect how humans and automated systems are able to transform, fulfill, and return products and services.*

A shared understanding of the ontology of supply chain management and military logistics is essential to highlight the equal importance of all components of the supply chain as a system and provides a more nuanced perspective on how the supply chain behaves as a complex adaptive system based on how the agents and processes dynamically interact. Umbrella terms like supply chain and sustainment do not suggest centralization nor bureaucratic 'empire-building antics.'[59] Instead, they unify organizations in purpose while enabling each more independent action. As the nation-state and industry provide the means of war to the military, there must be a common language between the parties. Commercial supply chain management and military logistics must have a common logic, although the grammar and the practice of military logistics/sustainment are distinct for the profession of arms.

Conceptualizing the supply chain environment as an aggregate provides a foundational theoretical principle in understanding it as both a military domain-equivalent and a complex adaptive system so that military commanders may identify points of leverage to create competitive advantages. Aggregation is a basic property of complex adaptive systems for people to identify larger patterns of behavior from lower hierarchies of less-complex interactions, similar to how an ant colony seems orderly at a macro level but chaotic at a micro level.[60] Aggregation allows the construction of simpler models to bring meaning to complexity and to provide the basis for a common vocabulary and structure to share that meaning widely, like international relations models. The aggregation property is also essential to construct useful linkages between the physical world of nature bound by physics and the cognitive space of human interaction at a scale beyond the control of any individual actor.

The 'economy' is an aggregation of a complex adaptive system. An economy is generally understood as an aggregate of conditions surrounding individual actors pertaining to physical geography and human interactions across production and consumption activities. An economy might be

---

59 Jeffrey C. Brlecic, "Logistics, CSS, Sustainment: Evolving Definitions of Support," *Army Sustainment* vol. 41, 5, (September-October 2009), 22, http://www.alu.army.mil/alog/2009/SepOct09/pdf/asust_septOct_09.pdf.
60 John H. Holland, *Hidden Order: How Adaptation Builds Complexity* (Helix Books, 1995), 11.

measured in aggregate as Gross Domestic Product to provide comparative insight into a country's economic health. As a neutral environment, a free-market economy has its own internal logic to seek new connections and find equilibrium across disparate agents with minimal intervention, metaphorically known as Adam Smith's "invisible hand." Conversely, the economy is an instrument of national power that can be a leverage point for a nation-state or transnational actor to exert control and influence over others and the broader environment.

However, sub-aggregates or intersectional aggregates may be needed to provide useful analysis of the economy to anticipate changes and adapt accordingly. This approach may yield counterintuitive logic about the complex system. Strategist and author Parag Khanna highlights this by intersecting geoeconomics with supply chain management principles, which he calls "connectography." He notes how increased global investment and supply chain interdependence since the fall of the Soviet Union have now made geopolitical rivals financially integrated like never before, making a state's effort to employ economic and military coercion much more difficult.[61] He stresses, "Supply chains thus diminish the incentives for conflict while decoupling from them raises the potential for antagonism to escalate."[62]

Khanna's theory of success for two entangled economic titans like the U.S. and China is not war but a 'tug-of-war' over factors in the supply chain environment, a paradox where "the longer it goes on, the more everyone wins."[63] In 2000, Craig Addison developed a similar supply chain deterrence theory on how Taiwan's semiconductor and information technology industry kept U.S. and China economies entangled and interdependent, thus making conflict too costly and preventing a cross-strait invasion by China.[64] The Silicone Shield theory has seen a resurgence with the rapid growth of the information technology industry, the People's Republic of China (PRC)'s aggression, and the Russian invasion of Ukraine.[65] Enmeshment of nation-states via supply chains creates an interdependency

---

61 Parag Khanna, *Connectography: Mapping the Future of Global Civilization* (New York: Random House, 2016), 149.
62 Ibid.,149.
63 Ibid.,150.
64 Craig Addison, "A 'Silicon Shield' Protects Taiwan from China" New York Times, September 29, 2000. https://www.nytimes.com/2000/09/29/opinion/IHT-a-silicon-shield-protects-taiwan-from-china.html.
65 Jared M. McKinney and Peter Harris, "Understanding the Deterrence Gap in the Taiwan Strait" War on the Rocks, February 12, 2024. https://warontherocks.com/2024/02/understanding-the-deterrence-gap-in-the-taiwan-strait/.

that raises the cost of decoupling or destruction from war too high and overtly aggressive actions too self-defeating. Therefore, national strategy on supply chain enmeshment should focus on derisking to ensure reserve capacity for disruptions by diversifying suppliers, protecting intellectual property, and building infrastructure.

The aggregation principle is already inherent within U.S. joint doctrine. The operational environment is "the *aggregate* of the conditions, circumstances, and influences that affect the employment of capabilities and bear on the decisions of the commander."[66] Furthermore, U.S. joint doctrine defines the information environment as "the *aggregate* of social, cultural, linguistic, psychological, technical, and physical factors that affect how humans and automated systems derive meaning from, act upon, and are impacted by information, including the individuals, organizations, and systems that collect, process, disseminate, or use information."[67] These abstractions may not be useful without intersecting the other relevant aggregates in Joint Doctrine and the emergent supply chain environment. The supply chain environment helps complete the picture of the surroundings that bear on the military commander's decision-making and employment of forces. It integrates with the recognized physical military domains of land, maritime, air, and space, the information environment (including cyberspace), and the electromagnetic environment.[68]

Network flows are also inherent in complex adaptive systems. At the network nodes (which can come and go) in the supply chain environment, agents direct the physical, informational, financial, and human capital flows to other agents in the system. This includes orchestrating systems pertaining to critical infrastructure, facilities/plants, manufacturing tools/technologies, data/intellectual property, software acquisition, digitization, legislation/regulations, financial/resourcing systems, workforce recruitment/development, partnerships, and setting conditions for innovation. The emphasis here is that supply chains involve various systems and activities that dynamically interact, can be resource intensive to acquire or develop, and are more difficult to synchronize.

Together, these systems can represent a state's organizational capital and financial capacity to adopt a new military innovation, thereby changing

---

66 Joint Chiefs of Staff. *Joint Warfighting*. Joint Publication 1, Volume 1 (Washington, DC: Joint Chiefs of Staff, August 27, 2023), I-5.
67 Ibid.
68 Ibid.

the balance of power in the international system. Dr. Michael Horowitz, Deputy Assistant Secretary of Defense for Force Development and Emerging Capabilities, presents a model where a state's organizational capacity has a greater impact on the rate and length of first-mover military innovation adoption advantage than financial ability or ability to fast-follow.[69] Rebalancing power through first-mover advantage provides leverage for credible deterrence or denial.

Organizational capital and financial capacity can also be envisioned as a technology factory for operational surprise. Hudson Institute fellows Brian Clark and Dan Patt suggest that persistent and overwhelming manufacturing of new military capabilities can dissuade an adversary from aggression and change behavior over time.[70] Center for Strategic and Budgetary Assessments Research Fellow Tyler Hacker presents alternatives to U.S. precision-guided munition stockpile shortages by leveraging digital engineering, advanced manufacturing, and modular ordnance systems to customize munitions for the optimal weapon-to-target pairings while dropping unit cost.[71] Investing in the underlying infrastructure of the supply chain and having the most talent in human capital and intellectual property sets conditions ripe for this kind of adaptability. Therefore, the supply chain environment is not just about having more weapon systems or a deeper arsenal to deter adversaries or engage in prolonged conflict—the supply chain environment is the engine of adaptation.

Reframing the supply chain from a linear function and material-centric activity to a multi-faceted, human-centered, and natural environmental factor of the battlespace is a better reflection of the reality in which supply chains are now so intertwined with the information environment, stretch globally across extended, hyperconnected networks, and behave as complex adaptive systems. In addition, it provides an intellectual framework for the military to operationalize this battlespace environmental factor more effectively across domains, across regions, across the spectrum of conflict,

---

69 Michael C. Horowitz, *The Diffusion of Military Power: Causes and Consequences for International Politics* (Princeton University Press: 2010), 48-49.
70 Clark, Brian, and Dan Patt. Campaigning to Dissuade: Applying Emerging Technologies to Engage and Succeed in the Information Age Security Competition. Washington, DC: Hudson Institute, July 2023. https://www.hudson.org/defense-strategy/campaigning-dissuade-applying-emerging-technologies-engage-succeed-information-age-bryan-clark-dan-patt.
71 Hacker, Tyler. Beyond Precision: Maintaining America's Stike Advantage in Great Power Conflict. Washington, DC: Center for Strategic and Budgetary Assessments, 2023. https://csbaonline.org/uploads/documents/Beyond_Precision_Report_CSBA8355_FINAL_web.pdf.

across the U.S. government, and with allies and partners, including industry, to achieve integrated deterrence.[72]

As Milan Vego exclaimed in criticizing early defunct airpower theories and net-centric warfare at the turn of the 21st century, inventing a domain or sphere of conflict with the ignorance of the nature of war, an ahistorical perspective, or the glorification of technology above all else is an artificial theory that is doomed to fail.[73] This proposal is not meant to label supply chains as a new physical domain of warfare to elicit short-term attention and confuse long-term implementation, as the term contested logistics has already done. Rather, the supply chain environment helps complete the picture of the surroundings that bear on the military commander's decision-making and employment of forces—the operational environment.[74] The supply chain environment integrates with the recognized physical military domains of land, maritime, air, and space, the information environment (including cyberspace), and the electromagnetic environment.[75]

An example is how the 2017 NotPetya cyberattack on global shipping leader A.P. Møller-Maersk, responsible for over 70 worldwide ports and about one-fifth of international trade, shut down business operations and inflicted an estimated two billion dollars in extended damages between Maersk and its logistics partners.[76] The virality of the NotPetya malware is astonishing as Maersk was an unintended target.[77] The NotPetya perpetrators, Russian military intelligence officers, were criminally charged by the U.S. Department of Justice in October 2020 for not only the damage done to the global transportation sector, but to the billion dollars in damages to Ukraine's energy grid in 2015 and 2016.[78] Maersk Line Limited

---

72 U.S. Department of Defense, *2022 National Defense Strategy*, (Washington, DC., October 2022), 1-2, https://media.defense.gov/2022/Oct/27/2003103845/-1/-1/1/2022-NATIONAL-DEFENSE-STRATEGY-NPR-MDR.PDF.
73 Milan Vego, "Domains of Conflict Versus the Art of War," in *Joint Operational Warfare: Theory and Practice*, Reprint of 1st ed. (Newport, RI: U.S. Naval War College, 2009), XIII-29.
74 Joint Chiefs of Staff. *Joint Warfighting*. Joint Publication 1, Volume 1 (Washington, DC: Joint Chiefs of Staff, August 27, 2023), I-5.
75 Ibid.
76 Daniel E. Capano, "Throwback Attack: How NotPetya accidentally took down global shipping giant Maersk," *Control Engineering* 70, no. 4 (May 2023), 39-41, https://www.proquest.com/trade-journals/throwback-attack-how-notpetya-accidentally-took/docview/2818731837/se-2.
77 Ibid.
78 Remarks By Assistant Attorney General for National Security John C. Demers on Announcement of Charges Against Russian Military Intelligence Officers, speech, Washington, DC, October 19, 2020. https://www.justice.gov/opa/speech/remarks-assistant-attorney-general-national-security-john-c-demers-announcement-charges.

President William Woodhour stated that in about half an hour, over 2,500 servers were shut down, putting the business in a communication blackout.[79] At the same public forum with Mr. Woodhour, the commander of U.S. Transportation Command (USTRANSCOM), General Darren McDew, USAF, warned about future transboundary threats, global supply chain operations being contested, and that USTRANSCOM was now leading the discussions on cyber in the DoD.[80] The cyberattack on Maersk demonstrates how businesses, as part of extended supply chains, are on the front lines of the new multi-dimensional battlespace and require a new level of interagency, international, and partner coordination.

The distinction between environment and domain must be clear. First, environments are inseparable from physical domains although they have a life of their own that cannot go unnoticed and thus must be elevated in importance to the equivalency of physical domains of warfare. As the NotPetya attack demonstrates, environments, like supply chains, are also different from the physical domains of warfare because they are a complex adaptive system characterized by the interconnectedness of many parts of both human and natural construct. In the case of supply chains, the natural construct is the land, sea, air, and space upon which goods are sourced, made, and delivered, making supply chains inherently multi-domain. Supply chains as complex adaptive systems also exhibit nonlinear behavior, where small changes in one area can have an amplified effect in a seemingly wholly unrelated area. Thus, a supply chain environment cannot be isolated or controlled like a physical domain. Different strategies and tactics are necessary.

As Air Force strategist Michael P. Kreuzer suggests with cyberspace and the information environment, the proposed supply chain environment theory is analogous to the physical domains of warfare, a unique environmental factor of the battlespace with a different logic that informs how it may genuinely be operationalized from the strategic to tactical level, integrate with multi-domain operations, and drive organizational change.[81] Furthermore, recognition of the supply chain environment is

---

79 Seafarers Log, "Gen. McDew is 'Huge Advocate' for Maritime," news release, December 1, 2017, https://www.seafarers.org/seafarerslogs/2017/12/gen-mcdew-is-huge-advocate-for-maritime/.
80 Ibid.
81 Michael Kreuzer, "Cyberspace is an Analogy, not a Domain: Rethinking Domains and Layers of Warfare for the Information Age," *Strategy Bridge* (July 8, 2021), https://thestrategybridge.org/the-bridge/2021/7/8/cyberspace-is-an-analogy-not-a-domain-rethinking-domains-and-layers-of-warfare-for-the-information-age.

similar to the development of 'information' as a doctrinal trinity: a form of national and military power, a unique aspect of the strategic and operational environments, and a joint warfighting function, because it can be an instrument of coercion and deterrence from the strategic to the tactical level.[82]

A holistic perspective of the supply chain environment from a domain lens can advance military theory, push the boundaries of strategy and operational art, and challenge public and professional thinking, like the approach by seapower theorist Alfred Thayer Mahan in developing a military theory distinct, yet additive, from the land domain, nested with the national power brought by dominion over the sea.[83] Certainly, Mahan understood the essence of geoeconomics and supply chains before those terms came into being, as his six elements of sea power capture both concepts, his virtuous cycle of deterrence links trade and finance with shipbuilding and a forward navy, his way of war intertwined commerce and naval strength as dual-offensive capabilities, and his conception of the fleet's defensive strength and staying power was through a network of strategic bases/ports.[84] If areas of commerce remain a place of competition, strategic signaling, conflict, and source of military capacity, then the battlespace cannot be visualized in a meaningful way when global commerce has converged so tightly with all the current joint doctrinal aspects of the operational environment, not just the maritime domain.

A declaration of the supply chain environment is akin to how the DoD declared in 2020 a transformation of its approach to the space domain from a support function in practice to a true warfighting domain coupled with the introduction of spacepower.[85] The changing domain required renewed focus to "drive enterprise-wide changes to policies, strategies, operations, investments, capabilities, and expertise for a new strategic environment" as the domain became more congested and contested.[86]

---

82 U.S. Marine Corps, *Information in Marine Corps Operations*, MCWP 8-10. (Washington, DC: Headquarters U.S. Marine Corps, February 29, 2024), 1-2 to 1-4, https://www.marines.mil/News/Publications/MCPEL/Electronic-Library-Display/Article/3712860/mcwp-8-10/.
83 Kevin D. McCranie, *Mahan, Corbett, and the Foundations of Naval Strategic Thought* (Annapolis, MD: Naval Institute Press, 2021), 40-41.
84 Kevin D. McCranie, 15, 22-28, 132-134, and 215-217.
85 U.S. Department of Defense, *Defense Space Strategy Summary*, (Washington, DC., June 2020), 6, https://media.defense.gov/2020/Jun/17/2002317391/-1/-1/1/2020_DEFENSE_SPACE_STRATEGY_SUMMARY.PDF.
86 U.S. Department of Defense, *Defense Space Strategy Summary*, (Washington, DC., June 2020), 1, https://media.defense.gov/2020/Jun/17/2002317391/-1/-1/1/2020_DEFENSE_SPACE_STRATEGY_SUMMARY.PDF.

Conclusion

It is a mental leap to consider the supply chain as a domain-equivalent aspect of the strategic and operational environments; however, that necessary step can unshackle the military from its dominant logic to think, train, and fight differently to meet the demands of future conflict. First, the hyperconnectivity brought by globalization and digitization, coupled with the changing character of war, have evolved the supply chain environment into its contemporary and complex form, a global competitive space spanning all physical domains of land, maritime, air, and space, along with cyberspace. A working definition of the supply chain environment *is the aggregate of material, informational, financial, and human capital factors that affect how humans and automated systems are able to transform, fulfill, and return products and services.*

Second, the supply chain environment can be leveraged as a source of power that provides nation-states, non-state actors, and militaries the capability and capacity to control or influence the behavior of others, particularly as it is the space where things are made, in which technology is invented, and innovations diffused. That combination stands to thwart or at least limit conflict in the future. Again, as stressed in this analysis, the ability to prepare for war and advance peace simultaneously, and as the same measure, is an infrequent but coveted moment in war; we are there now.

Third, the supply chain environment is a physical and cognitive maneuver space where military advantages can be gained or lost, thus making it imperative for the military commander, not just staff logisticians, to understand and assert leadership. Fourth, the buzzword term contested logistics must be abandoned as logistics is simply a function, primarily military in context, and the supply chain environment in which that function occurs is the thing that is contested across the competition continuum. This perspective means the civilian side of the equation gets its full attention. Again, war can be limited and threatened as a state's most crucial function or act.

11

Artificial Intelligence and Logistics on the Modern Battlefield

Stacy Tomic, Michael Posey, and Paul Lushenko

Scholars have explored drones' proliferation, effectiveness, and normative implications, particularly high-tier armed and networked drones like the U.S.-manufactured MQ-9 Reaper. Extant research has not yet explored the implications of low-tier and mid-tier drones as much, including those enhanced with Artificial Intelligence (AI), largely because these capabilities are rapidly developing.[1] Nevertheless, policymakers, practitioners, and war theorists agree that AI-enhanced capabilities constitute a shift in the character of war and across all tactical, operational, and strategic levels. In this case, observers often privilege a narrow rather than generative form of AI. The latter consists of generative pre-trained transformers, such as those used in Large Language Models like ChatGPT, that stack algorithms to create artificial neural networks that improve probabilistic reasoning to classify objects and forecast outcomes based on representative data.[2] Narrow AI, on the other hand, is function specific. In other words, it is deliberately designed for constrained tasks such as optimizing targeting workflows based on algorithms trained on representative data.[3]

Given this context, this chapter aims to advance our understanding of the implications of narrow AI on sustainment operations, given broader shifts in the trajectory of future war considering the emergence of AI. We argue that while AI-enhanced capabilities can provide efficiencies for sustainment operations at the tactical level, the complexity imposed by

---
1 Dominika Kunertova, "Drones Have Boots: Learning from Russia's War in Ukraine," *Contemporary Security Policy*, Vol. 44, No. 4 (2023), 576-591.
2 Ash Rossiter and Peter Layton, *Warfare in the Robotics Age*. (Boulder: Lynne Rienner Publishers, 2024), 62-66.
3 Michael C. Horowitz, "Artificial Intelligence, International Competition, and the Balance of Power," *Texas National Security Review*, Vol. 1, No. 3 (2018), 36-57.

different strategic contexts of conflict, ranging from humanitarian assistance and disaster relief to large-scale ground combat operations, imposes the requirement for more, not less, human oversight of sustainment operations globally, regionally, and in different theaters of operation. The tactical level of war refers to battlefield actions such as patrols or raids; the operational level of war characterizes a military's synchronization of tactical actions to achieve broader military objectives, such as destroying an opponent's army; and the strategic level of war characterizes the way these military objectives combine to achieve overall political aims, such as war termination.

Based on this argument, the remainder of this chapter unfolds in four parts. First, we introduce a conceptual framework to understand the potential implications of AI-enhanced capabilities on future war. We then connect this framework to the principles of logistics drawn from multinational, joint, and service-oriented (U.S. Army) doctrine. Doing so enables us to forecast the implications of AI for battlefield logistics, which we do in the third section. We conclude by discussing the policy, modernization, and research implications of our analysis.

AI and the Shifting Character of War

Militaries can use AI to inform various decisions with varying degrees of human oversight. In terms of decision-making, militaries can optimize algorithms to perform tactical operations. They can also design them to conduct strategic deliberations in support of overall war aims.

Tactically, AI can enhance the lethality of field commanders by rapidly analyzing large quantities of data drawn from sensors distributed across the battlefield to generate targeting options faster than adversaries.[4] By doing so, AI can shorten the "sensor-to-shooter" timeline, which refers to the interval of time between acquiring and prosecuting a target. AI expert Mike Drennan notes that "a robot evaluates millions of possible scenarios and selects the decision that produces the best outcome. This level of automation can potentially mitigate risk by quickly and decisively waging war."[5] Drawing from Google's analytical techniques, Project Maven attempts to rapidly categorize suspected targets' behaviors and artifacts, meaning what they do and carry, in terms of 'hostile intent' or 'hostile act,'

---

[4] Jon R. Lindsay, "War is From Mars, AI is from Venus: Rediscovering the Institutional Context of Military Automation," *Texas National Security Review*, Vol. 7, No. 1 (2023), 1-33.
[5] James E. Drennan, "How to Fight an Unmanned War," in *The U.S. Naval Institute on Naval Innovation*, Ed. John E. Jackson, (Annapolis: Naval Institute Press, 2015), 118.

to assist U.S. military commanders in making targeting decisions.[6] Strategically, AI can also assist planners in synchronizing objectives with different warfighting approaches and finite resources regarding[7] both materiel and personnel. New AI-enabled capabilities might emerge that would replace humans during future operations, including crafting strategic direction and national-level strategies.[8]

In crafting such strategic guidance, countries can also calibrate the amount of human oversight of AI. These technologies can be designed to allow for greater degrees of human oversight, ranging from human 'on-the-loop' to mixed-initiative control, affording human operators the ability to control their agency over decision-making, much like setting a home's temperature rheostat.[9] Ashby once referred to using computers with oversight as "intelligence amplification," wherein information technologies have the potential to augment human intelligence in support of rapid knowledge production.[10] Systems designed this way are also called semi-autonomous, such that humans ultimately remain in charge. This human oversight characterizes how most AI-enhanced weapons systems, including the MQ-9 Reaper drone, the most advanced drone in the world, currently operate.

However, Drennan notes that this approach may impose a tradeoff between decision-making and time management, wherein human agency is retained at the cost of "the advantage of near-instantaneous decision-making."[11] Thus, countries can also design AI-enhanced military technologies with less human oversight to achieve the intent of near-instantaneous decision-making. Experts often refer to these systems as killer robots because the human is 'off-the-loop.'[12]

As reflected in Figure 11.1, variations in the decision-making level and extent of human oversight suggest four types of AI-enabled warfare

---

6 Taylor K. Woodcock, "Human/Machine(-Learning) Interactions, Human Agency and the International Humanitarian Law Proportionality Standard," *Global Society*, Vol. 38, No. 1 (2023), 1-22.
7 Anthony King, "Digital Targeting: Artificial Intelligence, Data, and Military Intelligence," *Journal of Global Security Studies*, Vol. 9, No. 2 (2024), 1-16.
8 Michael Mayer, "Trusting machine intelligence: artificial intelligence and human-autonomy teaming in military operations," *Defense & Security Analysis*, Vol. 39, No. 4 (2023), 521-538.
9 Avi Goldfarb and Jon R. Lindsay, "Prediction and Judgment: Why Artificial Intelligence Increases the Importance of Humans in War," *International Security*, Vol. 46, No. 3 (2022), 7-50.
10 W. Ross Ashby, *An Introduction to Cybernetics*. (London: Chapman & Hall, 1956).
11 Drennan, "How to Fight an Unmanned War," 122.
12 Paul Scharre, "Debunking the AI Arms Race Theory," *Texas National Security Review*, Vol. 4, No. 3 (2021): 121-132.

that different countries might pursue using AI. First, countries could use AI for tactical decision-making with human oversight. This type is what autonomous weapons expert Paul Scharre calls "centaur warfighting."[13] This model of AI-enabled warfare is named after a creature from Greek mythology with the upper body of a human and the lower body and legs of a horse. Centaur warfare emphasizes human control of machines for battlefield purposes, such as destroying an enemy's arms cache.

Countries could also use AI for tactical decision-making with little, if any, human oversight. This approach flips centaur warfare on its head, evoking another mythical creature from ancient Greece—the minotaur. Unlike the centaur, the minotaur has the head and tail of a bull and the body of a man. "Minotaur warfare," according to philosophers Robert Sparrow and Henschke, is characterized by machine control of humans during combat and across warfighting domains.[14] This type can range from machine control of patrols of soldiers on the ground to constellations of warships on the ocean to formations of bombers in the air to groupings of satellites in space.

Third, countries could adopt strategic decision-making through an "AI-general," which University of Leicester Researcher Cameron Hunter and Professor Bleddyn Bowen characterize as a "singleton" type of AI-enabled warfare.[15] This approach invests AI with extraordinary latitude to shape countries' warfighting strategy trajectories, constituting an extreme form of minotaur warfare. The AI-general model of warfare could allow countries, if not nonstate groups such as terrorists, to gain and maintain advantages over adversaries in time and space that shape overall war outcomes. It may even have severe implications for the offense-defense balance between countries during conflict,[16] perhaps including serving to "save humanity before destroying it."[17]

---

13 Paul Scharre, "Centaur Warfighting: The False Choice of Humans vs. Automation," *Temple International & Comparative Law Journal*, Vol. 30, No. 1 (2016), 151-165.
14 Robert J. Sparrow and Adam Henschke, "Minotaurs, Not Centaurs: The Future of Manned-Unmanned Teaming," *Parameters*, Vol. 53, No. 1 (2023), 115-130.
15 Cameron Hunter and Bleddyn E. Bown, "We'll Never Have a Model of an AI Major-General: Artificial Intelligence, Command Decisions, and Kitsch Visions of War," *Journal of Strategic Studies*, Vol. 27, No. 1 (2023), 1-31.
16 Ajey Lele, "Blockchain," in *Smart Innovation, Systems and Technologies*. (Singapore: Springer, 2019), 197-202.
17 Zachary Kallenborn, "Policy makers should plan for superintelligent AI, even if it never happens," *Bulletin of the Atomic Scientists*, Dec. 21, 2023.
https://thebulletin.org/2023/12/policy-makers-should-plan-for-superintelligent-ai-even-if-it-never-happens/#post-heading.

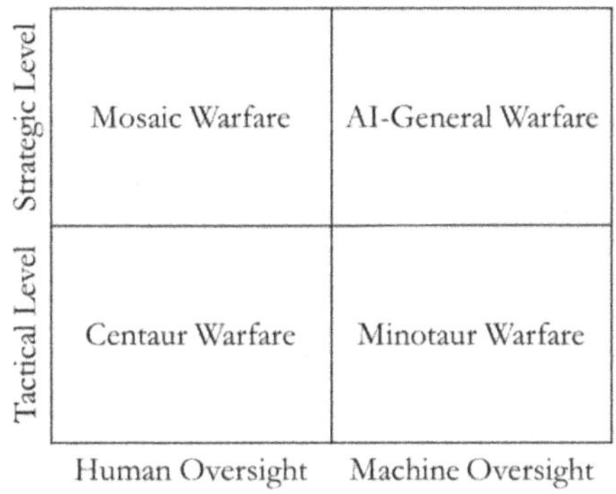

	Human Oversight	Machine Oversight
Strategic Level	Mosaic Warfare	AI-General Warfare
Tactical Level	Centaur Warfare	Minotaur Warfare

Figure 11.1: Patterns of AI-Enabled Warfare (figure created by Dr. Paul Lushenko)

Finally, "mosaic warfare" embraces the use of algorithms to facilitate timely and effective strategic decision-making but retains human oversight of AI.[18] The intent of this warfighting model—which US Marine Corps General (Retired) John Allen calls "hyperwar" and scholars often refer to as algorithmic decision-support systems—is two-fold.[19] First, mosaic warfare is designed to rapidly identify, impose, and exploit vulnerabilities against adversaries, especially in the context of great-power conflict shaped by near-peer militaries. Second, mosaic warfare also optimizes enabling tasks through emerging capabilities like COA-GPT, which predicts enemy courses of action, shapes strategy, and builds redundancy into operations, especially sustainment, since this enables power projection.[20]

---

[18] Timothy Grayson, "DARPA Tiles Together a Vision of Mosaic Warfare: Banking on cost-effective complexity to overwhelm adversaries," Defense Advanced Research Projects Agency, undated. https://www.darpa.mil/work-with-us/darpa-tiles-together-a-vision-of-mosiac-warfare.
[19] Amir Husain, John Rutherford Allen, Robert O. Work, August Cole, Paul Scharre, Bruce Porter, Wendy R. Anderson, and Jim Townsend, *Hyperwar: Conflict and Competition in the AI Century*. (Austin: SparkCognition Press, 2018); Woodcock, "Human/Machine(-Learning) Interactions, Human Agency and the International Humanitarian Law Proportionality Standard," 1-23.
[20] Vinicius G. Goecks and Nicholas Waytowich, "COA-GPT: Generative Pre-trained Transformers for Accelerated Course of Action Development in Military Operations," Cornell University, ArXiv, Feb. 1, 2024, last revised Mar. 28, 2024. https://arxiv.org/abs/2402.01786.

Logistics, Sustainment, and the Principles of Logistics

This section connects our typology of patterns of AI-enabled warfare to the requirements for logistics during future warfare, which are shaped by fundamental principles that govern logistical operations. We achieve this overall intent in three parts. First, we define logistics. Second, we define the relationship between logistics and sustainment operations. Third, we relate these definitions to the principles of logistics, which we use to inductively adjudicate the implications of particularly Centaur and Minotaur warfare for logistics on the future AI-enabled battlefield. In defining these critical terms, we draw from three multinational, joint, and service-oriented doctrine sources. These documents include North Atlantic Treaty Organization (NATO) Allied Joint Publication 4 (AJP-4)—*Allied Joint Doctrine for Logistics*; US Joint Publication 4-0 (JP 4-0)—*Joint Logistics*; and US Army Doctrine Publication 4-0 (ADP 4-0)—*Sustainment*.

Logistics

Allied Joint Doctrine for Logistics defines logistics as "the science of planning and carrying out the movement and maintenance of forces." It adds that, in the most comprehensive sense, "logistics refers to the aspects of military operations that deal with: design and development, acquisition, storage, movement, distribution, maintenance, evacuation and disposition of materiel; transport of personnel; acquisition, construction, maintenance, operation, and disposition of facilities; acquisition or furnishing of services; and medical and health service support."[21] Similarly, Joint Logistics states that "logistics includes planning, executing, and assessing the movement and support of forces."[22] This joint doctrine also identifies the core logistics functions as deployment and distribution, supply, maintenance, logistics services, operational contract support, engineering, and joint health services.[23] Finally, Sustainment defines logistics as the "planning and executing the movement and support of forces." This service-oriented document adds that logistics characterizes "design and development; acquisition, storage, movement, distribution, maintenance, and disposition of materiel; acquisition or construction, maintenance, operation, and disposition of facilities; and

---

21 North Atlantic Treaty Organization, *Allied Joint Doctrine for Logistics*, AJP-4, Edition B, Version 1. (Brussels: NATO Standardization Office, 2018), 1-1.
22 Joint Chiefs of Staff, *Joint Logistics*, JP 4-0. (Washington, DC: Joint Chiefs of Staff, 2023), I-1.
23 Ibid., II-1.

acquisition or furnishing of services." Moreover, *Sustainment* contends, "Army logistics elements are: maintenance, transportation, supply, field services, distribution, operational contract support, and general engineering."[24]

Sustainment

What is the relationship between logistics and sustainment operations? U.S. Army doctrine defines sustainment as one of the seven joint functions, as adumbrated in JP 4-0, or one of the six warfighting functions.[25] This service-oriented doctrine also identifies logistics as one of the four elements of the sustainment warfighting function. The remaining three elements include financial management, personnel services, and health service support.[26] On the other hand, *Joint Logistics* states that "sustainment frames both the objective (ends) and provides the means, in both capabilities and capacity, to achieve those ends. Logistics represents the activity of planning for and employing the capabilities within the capacity available to achieve the stated ends."[27] Contrary to service-oriented and joint doctrines, multinational doctrine is more ambiguous as NATO doctrine mainly treats sustainment as a national responsibility. The *Allied Joint Doctrine for Logistics* defines logistic sustainment as "the process and mechanism by which sustainability is achieved and which consists of supplying a force with consumables and replacing combat losses and non-combat attrition of equipment in order to maintain the force's combat power for the duration required to meet its objectives."[28]

We reconcile these doctrinal consistencies by offering a simplified definition of logistics and its relationship to sustainment. For this chapter, we define logistics as the planning, coordinating, and executing movement, supply, and maintenance of personnel, materiel, and equipment in support of military operations. We will also categorize logistics as one of the four elements of sustainment, similar to the U.S. joint doctrine.

---

24 U.S. Department of the Army, *Sustainment*, ADP 4-0. (Washington, DC: Department of the Army, 2019), 1-1.
25 JP 4-0, *Joint Logistics*, I-2; ADP 4-0, *Sustainment*, 1-1.
26 Ibid.
27 JP 4-0, *Joint Logistics*, I-2.
28 AJP-4, *Allied Joint Doctrine for Logistics*, Lexicon-5.

Principles of Logistics

Similar to competing but related definitions of logistics and sustainment, the principles of logistics vary across multinational, joint, and service-oriented doctrine. *Joint Logistics* states that "logisticians can use the principles of logistics as a guideline to assess how effectively logistics are integrated into plans and execution."[29] Similarly, *Sustainment* equates the principles of sustainment to logistics, arguing that they "are essential to maintaining combat power, enabling strategic and operational reach, and providing Army forces with endurance."[30] The principles of logistics, then, are necessary to ensure effective sustainment of military operations. As such, *Allied Joint Doctrine for Logistics* identifies ten principles of logistics, including national and collective responsibility for logistics, authority, primacy of operational requirements, cooperation and coordination, assured provision, sufficiency, efficiency, simplicity, flexibility, and visibility.[31] *Joint Logistics* identifies eight principles: resilience, responsiveness, feasibility, flexibility, simplicity, visibility, cooperation, and economy.[32] *Sustainment* also identifies eight principles, but these are slightly different, including integration, anticipation, responsiveness, simplicity, economy, survivability, continuity, and improvisation.[33]

Notwithstanding differences in the principles of logistics across these three source documents, we observe a degree of consistency in terms of anticipation, responsiveness, improvisation, simplicity, and economy. Anticipation constitutes "the ability to foresee operational requirements and initiate necessary actions that most appropriately satisfy a response without waiting for operations orders or fragmentary orders."[34] Responsiveness relates to "the ability to react to changing requirements and respond to meet the needs to maintain support."[35] Similar to flexibility, improvisation is consistent with "the ability to adapt sustainment operations to unexpected situations or circumstances affecting a mission."[36] Simplicity, akin to parsimony, "fosters efficiency in planning and execution and enables

---

29 JP 4-0, Joint Logistics, I-11.
30 ADP 4-0, *Sustainment*, 1-2.
31 AJP-4, *Allied Joint Doctrine for Logistics*, 1-1 to 1-3.
32 JP 4-0, *Joint Logistics*, I-11 to I-13.
33 ADP 4-0, *Sustainment*, 1-2 to 1-4.
34 ADP 4-0, Sustainment, 1-3.
35 Ibid.
36 Ibid., 1-4.

Artifical Intelligence and Logistics on the Modern Battlefield 177

	Principles of Logistics				
	Anticipation	Responsiveness	Improvisation	Simplicity	Economy
Centaur		✓	✓		
Minotaur	✓			✓	✓

Figure 11.2: AI-Enabled Warfare and the Principles of Logistics (figure created by Dr. Paul Lushenko)

more effective control over logistics activities and operations."[37] Economy, or optimization, ensures "the minimum number of resources required to achieve a specific objective."[38]

These principles of logistics, distilled from across multinational, joint, and service-oriented doctrine, are helpful to inductively analyze the implications of AI-enabled logistics at the tactical level of war, which relate to our models of centaur and minotaur AI-enabled warfare. In the next section, we draw on Figure 11.2 to illustrate the ability of the centaur and minotaur models of AI-enabled warfare to meet the principles of logistics, which is informed by our experience, expertise, and research method. Specifically, we use a method of counterfactual analysis, wherein we draw on past—and anticipated—conflicts to vary some aspects of logistical operations at the operational and tactical levels of war.[39] This methodology allows us to provide at least a provisional assessment of the conditions under which AI will likely shape future battlefield logistics, as reflected in Figure 11.2.

Logistics on the Future AI-Enabled Battlefield

This section examines the connection between the patterns of AI-enabled warfare at the tactical level of war and the principles of logistics. To scope our analysis, we focus on four elements of logistics, including fuel (Class III) resupply, ammunition (Class V) resupply, medical operations (personnel treatment and evacuation), and maintenance operations (equipment repair and evacuation). Army logisticians frequently refer to these four elements as '35MM,' which are essential—decisive, even—to sustain combat

---

37 JP 4-0, *Joint Logistics*, I-12.
38 Ibid., I-13.
39 Richard Ned Lebow, *Forbidden Fruit: Counterfactuals and International Relation*. (Princeton: Princeton University Press, 2010).

operations. For the patterns of centaur and minotaur AI-enabled warfare, we will first discuss their alignment with the principles of logistics. Next, we will highlight how logisticians can best capitalize on these models of AI-enabled war.

Centaur Warfare

Frequently, because of the complexity of a situation, a human must ultimately decide how to execute the sustainment joint function. Many situations a commander or senior logistics planner will face are systematic combinations of many disparate disciplines and areas of expertise. A snapshot understanding of the battlefield—what the great Prussian military theorist Carl von Clausewitz called *"coup d'oeil"*—is often necessary for a general officer. Clausewitz defined *coup d'oeil* as "the rapid discovery of a truth which to the ordinary mind is either not visible at all or only becomes so after long examination and reflection."[40] Some warfighters see this as the commander's intuition, a combination of expertise, experience, and diverse career paths. Scholars seeking to define how the reasoning comes about may call it 'abduction,' a type of inference that seeks the best explanation given the available information.[41]

While there are rapid advances in AI, humans are still the best at abductive reasoning, especially in complex, dynamic environments.[42] To have this sort of reasoning, AI must be cross-disciplined and able to think as a human, often referred to a General or 'strong' AI, which remains largely aspirational.[43] Narrow or 'weak' AI, designed to focus on performing a single, constrained task, is the most proven and common form of AI.[44] In

---

40 Carl von Clausewitz, *On War*, Ed. Anatol Rapoport. (1832; repr., London: Penguin, 1968), 141.
41 Igor Douven, "Abduction," in *The Stanford Encyclopedia of Philosophy*, Summer 2021 Edition, Ed. Edward N. Zalta, The Metaphysics Research Lab, Center for the Study of Language and Information. (Stanford: Stanford University, 2021).
https://plato.stanford.edu/archives/sum2021/entries/abduction/.
42 Ben Dickson, "Abductive Inference: The Blind Spot of Artificial Intelligence," *TechTalks*, Sep. 20, 2021. https://bdtechtalks.com/2021/09/20/myth-of-artificial-intelligence-erik-larson/.
43 Ibid. See also Marco Iansiti and Karim R. Lakhani, "Competing in the Age of AI," in *HBR's 10 Must Reads on AI (with bonus article "How to Win with Machine Learning" by Ajay Agrawal, Joshua Gans, and Avi Goldfarb).* (Cambridge: Harvard University Press, 2023), 1-14.
44 "Narrow AI vs General AI and Super AI: How Do They Differ?" How to Learn Machine Learning: Your repository of resources to learn Machine Learning. Apr. 23, 2023.
https://howtolearnmachinelearning.com/articles/narrow-ai-vs-general-ai/. See also Iansiti and Lakhani, "Competing in the Age of AI."

these complex, dynamic situations, when humans must ultimately decide how to respond promptly to a rapidly evolving environment, the centaur model provides an avenue for logistics planners.

As a tool for logistics planners, the centaur model works best for sustainment in four different types of environments. First, centaur warfare relates to event-drive or dynamic situations, such as providing sustainment for a large natural disaster, unanticipated maintenance issues, or executing a medical evacuation. Second, centaur warfare is relevant when risk-to-force is substantially high, like combat search and rescue operations, conducting sustainment before battle damage assessment is complete, or when resupply endangers several aspects of the supply chain. Third, centaur warfare is best for situations presenting significant moral ambiguity, such as when a decision to resupply Unit A or Unit B will result in the inevitable death of one of the units. Finally, centaur warfare is proper when there is a need for increased contextual understanding, such as weighing the value of an emergency resupply of Class III or V for an encircled unit against the value of a certain number of casualties expected to occur during the resupply effort or understanding how to conduct sustainment efforts with cultural sensitivity. These four situations where a human should be the decider with machine input all call for the logistics principles of responsiveness and improvisation.

Responsiveness. Responsiveness requires flexibility and a keen understanding of sustainment needs, sustainment assets, logistics capabilities, and employment possibilities. Lieutenant General William "Gus" Pagonis exemplified the principle of responsiveness during the early 1990s in Operations Desert Shield and Desert Storm. For instance, he challenged the 22nd Support Command to develop theater-wide support for U.S. and Allied reception, staging, onward movement, and integration with Army forces arriving in Saudi Arabia.[45] Further, Pagonis established the 22nd Theater Army Area Command (22nd TAACOM) to support Coalition forces in the AOR shortly after U.S. forces arrived in the Central Command theater. Additionally, Pagonis ensured his logistics command rapidly responded to assistance from the Kingdom of Saudi Arabia by crafting contracts for food, fuel, water, and transport.[46]

Pagonis also enabled the rapid logistical movement to support General Norman Schwartzkopf's 'left hook' during the first 21 days of the

---

45 William G. Pagonis and Michael D. Krause, "Theater Logistics in the Gulf War," *Army Logistician*, PB 700-92-4 (July-August 1992), 3.
46 Ibid.

air campaign, when Iraqi forces did not have their command-and-control structures (for example, its 'eyes and ears') intact.[47] Pagonis demonstrated responsiveness by understanding Schwartzkopf's ground offensive, synchronizing logistical support, and delivering supplies where they were required. With AI available, some of the tasks the 22nd TAACOM faced could be automated, but the significant decisions about how best to respond to the chaos of moving hundreds of thousands of troops into Saudi Arabia while simultaneously ensuring logistics would coincide with the ground commander's offensive would still require savvy human thinking.

Although this example highlights the necessary human-led responsiveness at the operational level of war, the same idea is also applicable at the tactical level of war. The results of armed conflict events are often highly unpredictable, and as such, the necessity lies with the human decision-maker to be responsive to those actions. Imagine that an adversary attacks a supply convoy, resulting in severe soldier injuries. Soldiers in the convoy would need to request medical evacuation (MEDEVAC) support. That 9-Line MEDEVAC request includes information like location, patient and injury information, security, and the terrain. With that information, the MEDEVAC units determine the quickest and best method to evacuate the soldiers for treatment. In this example, humans, capable of abductive reasoning, must make the most responsive decision.

Improvisation. Synchronizing logistics with forces in austere combat environments requires both responsiveness and improvisation. Logistics planners in Afghanistan had to employ abductive reasoning for the *"coup d'oeil"* needed to understand what was happening. For instance, logistics planners demonstrated improvisation when establishing the Northern Distribution Network (NDN) to provide alternative supply routes to Afghanistan's sole ground line of communication, which came from Pakistan. Logistics planners in 2009 and 2010 worked alongside diplomats as they sought to bring fuel, electricity, and supplies to U.S. and coalition forces via the Central Asian states.[48] With tenuous agreements, many logistical movements relied on partners who could withdraw their support

---

47 Ibid., 6.
48 Kelly J. Lawler, "Learning from Northern Distribution Network Operations," U.S. Army News. Jul. 9, 2014. https://www.army.mil/article/128738/learning_from_northern_distribution_network_operations.

for the NDN anytime.[49] Further, the NDN's logistical success relied on novel supply routes via Latvia, Russia, Georgia, Azerbaijan, Kazakhstan, Kyrgyzstan, Tajikistan, and Uzbekistan.[50] The combination of air, sea, and land routes these logistics planners developed required understanding the diplomatic nuances of what supplies logisticians could move, what in mode, and through what country. While AI could assist with the details, human planners needed to drive and implement this demonstration of the logistical principle of improvisation to resupply Afghanistan from the north.

What does improvisation look like at the tactical level of war? Actions in war rarely go as perfectly as planned. For instance, imagine a battle-damaged tank that slides into a ditch. Mechanics driving an M88 armored recovery vehicle can pull the tank out of the ditch, then using improvisation, jerry-rig the steering mechanism with some parts on hand. Maintenance personnel would use non-standard repair techniques to ensure the tank is functional enough to return to battle. In sum, while machines can help a commander or logistics planner comprehend what is happening, the logistics principles of responsiveness and improvisation are uniquely suited to the centaur's human-led, machine-assisted methods.

Minotaur Warfare

The minotaur model is generally better when human decision-making is not required for anything other than a 'sanity check.' For example, this model works well when logistics are predictive or when they can be deliberately planned. Additionally, when the risk to friendly forces is lower, there is less moral ambiguity in employing the minotaur model. Therefore, logisticians should use this model for routine sustainment decisions to determine feasibility or when ethical decisions are required. Overall, the minotaur model is a viable tool for logisticians when the situation dictates less need for contextual understanding, and rapidity, efficiency, and economy are operationally advantageous. As such, the minotaur model can facilitate the achievement of three logistics principles: anticipation, simplicity, and economy.

---

49 Andrew C. Kuchins, Daniel Kimmage, Joseph Ferguson, Thomas M. Sanderson, Alexandros Petersen, Heidi Hoogerbeets, and David Gordon, "The Northern Distribution Network and Afghanistan," Center for Strategic and International Studies. Jan. 6, 2010. https://www.csis.org/analysis/northern-distribution-network-and-afghanistan.
50 Andrew P. Betson, "Nothing is Simple in Afghanistan: The Principles of Sustainment and Logistics in Alexander's Shadow," *Military Review*, Vol. 92, No. 5 (2012), 56.

Anticipation. Anticipation requires the logistics commander's plan to be synchronized with the maneuver commander's, especially in ambiguous operational environments. One way that commanders forecast how to execute operations during uncertain situations is by leveraging decision points. Decision points are "points in space and at the latest time when the commander or staff anticipates making a key decision concerning a specific course of action."[51] A great example of a logistics staff anticipating their maneuver commander's decision points is Lieutenant General George Patton's Third Army in December 1944, when the force changed its axis of advance.[52] Patton's sharp 90-degree turn enabled Third Army to counterattack the Germans along their southern flank and relieve isolated forces at Bastogne.[53] Because the Third Army logisticians had set the conditions for the operational maneuver, General Dwight Eisenhower quickly approved the maneuver when Patton briefed him on it.[54]

Quick, informed decisions remain essential to success in war. In contemporary warfare, characterized by a proliferation of sensors across the battlefield, AI provides a valuable tool to aid logisticians in anticipating their commander's needs. In particular, the minotaur model can enable faster and more complete anticipation in logistics. As a planning group develops decision points for the commander, the logistics planner can use AI to forecast the logistical details required to support the commander's decision. With some human input, the machine can synthesize how logisticians would best move various classes of resupply, how to protect the supply lines, and whether a commander's decision is feasible regarding sustainment.

Imagine a division involved in large-scale combat operations that reaches a pre-planned operational decision point. Being well prepared, the logistics planners use predictive analytics to help them anticipate their commander's needs. By forecasting the requisite 35MM for the division commander during three different courses of action (COAs), logistics

---

51 Joint Chiefs of Staff, *Joint Planning*, JP 5-0. (Washington, DC: Joint Chiefs of Staff, 2020), GL-7.
52 Doug Chalmers and Craig A. Falk, "Forgotten Basics That Enable Decisive Action," U.S. Army News. Apr. 19, 2018. https://www.army.mil/article/203906/forgotten_basics_that_enable_decisive_action.
53 History. "General Patton Relieves Allies at Bastogne," This Day in History, December 26. Nov. 16, 2009. Last Updated Dec. 21, 2023. https://www.history.com/this-day-in-history/patton-relieves-bastogne.
54 Doug Chalmers and Craig A. Falk, "Forgotten Basics That Enable Decisive Action," U.S. Army News. Apr. 19, 2018. https://www.army.mil/article/203906/forgotten_basics_that_enable_decisive_action.

planners can help the commander make a sound decision. If the first COA cannot provide the needed medical support, this COA may no longer be feasible. Similarly, if the second COA strains the fuel supply beyond its limits, the COA incurs an increased risk to the mission. Finally, if the third COA can meet all the 35MM requirements, that consideration will help the commander make an operationally advantageous decision. In this way, minotaur warfare can leverage the power of machine learning to forecast resupply for the division's anticipated requirements at a decision point.

Simplicity. Simplicity in logistics enables efficiency and more effective command and control, commonly referred to as "C2." In 2018, a few years before the most recent boom in AI, when writing about U.S. Army logistics in the Korean War, historian Dr. James A. Huston stated: "In the planning process, certain data must be gathered and evaluated, procedures considered, limitations studied, and assets analyzed. This makes the actual support simpler, quicker, and more efficient when the necessity arises. New data may be used more quickly and effectively if only the basic questions have been sought out in advance."[55] Huston describes a link between AI-enabled data analytics and the simplicity principle of logistics. Humans do not need to be involved in routine decisions where demand is consistent. For example, during steady state operations with routine 35MM requirements, machines are better at computing the scheduled resupply of class III or V and scheduling maintenance intervals using predictive maintenance. By doing so, logistical planners can focus on more complex decisions. With little human interaction, the AI minotaur model can aid logistics efficiency and thereby enable simplicity.

Economy. The principle of economy in logistics means planners must reduce waste and increase efficiency. Often, simplicity and economy are correlated. One method to improve efficiency is to forecast logistics accurately. In their piece on tactical logistics, two logistics officers, Michael Johnson and Brent Coryell, note that "relying on a default push of supplies results in wasted manhours, increases the risk to Soldiers, and commits unneeded logistic assets."[56] Their observations suggest that not forecasting can cause even more problems at the operational level, leading to a reduced logistical economy. As such, leveraging trained AI algorithms to forecast operational sustainment demands will reduce costs and increase efficiencies,

---

55 James A. Huston, "Korea and Logistics," in *The Long Haul: Sustainment Operations in LSCO*, Ed. Keith R. Beurskens. (Fort Leavenworth, KS: Army University Press, 2018), 82.
56 Michael Johnson and Brent A. Coryell. "Logistics Forecasting and Estimates in the Brigade Combat Team," *Armor: Mounted Maneuver Journal*, Vol. 127, No. 4 (2016).

exemplifying the economy's logistical principle. For instance, logisticians can use AI to optimize inventory management to ensure the maintenance of optimum stockage levels based on mission requirements, especially for items like class IX (repair parts) and class VIII (medical supplies).

Implications

Our analysis above constitutes a first but important plausibility probe regarding the implications of AI during future conflict, particularly for battlefield logistics. Our analysis suggests that whereas AI-enabled decision-making can both optimize and simplify battlefield logistics and help anticipate key decisions for routine tasks, soldiers and commanders, given their abduction, are still postured to better respond to logistical requirements, especially given their ingenuity. In other words, logistics on the future AI-enabled battlefield requires more, not less, human oversight. Indeed, war is fraught with rampant uncertainty and chronic misinformation, the proverbial 'fog and friction,' which AI will exacerbate due to a deluge of information. This conclusion has several policy, modernization, and research implications.

First, though the U.S. Army has experimented with AI-enabled workflows, particularly for targeting in the context of large-scale ground combat operations, this approach is not standardized across the force and calibrated at different echelons. In the future, senior leaders should establish policies to aggregate pockets of innovation across the U.S. Army for AI-enabled logistics, drawing from the experiences and insights from the 18th Airborne Corps. Similarly, U.S. political and military officials emphasize interoperability with multinational partners, suggesting that this offers a comparative advantage during a great-power war with China or Russia. According to the American *National Security Strategy*, China and Russia are America's respective 'pacing' and 'acute' threats. Nevertheless, AI will exacerbate partnered military operations, which are already complicated by country-level caveats restricting troop contributions and operations.[57] As such, U.S. Army Senior Leaders, in concert with the Joint Force and Department of Defense, should promulgate standards that govern

---

57 Stephen M. Saideman and David P. Auerswald, "Comparing Caveats: Understanding the Sources of National Restrictions upon NATO's Mission in Afghanistan," *International Studies Quarterly*, Vol. 56, No. 1 (2012), 67–84; Erik Lin-Greenberg, "Allies and Artificial Intelligence: Obstacles to Operations and Decision-Making," *Texas National Security Review*, Vol. 3, No. 2 (2020), 56-76.

the integration of AI into coalition operations, particularly regarding data, access privileges, and classification.

Second, AI-enabled logistics is a promising contribution to the emerging Joint Warfighting Concept, namely the key tenant of resilient logistics.[58] Thus, planners and logisticians at the Army Service Component Commands, Theater Special Operations Commands, and Combatant Commands should draw from our analysis to determine the implications for campaign planning. The Indo-Asia-Pacific region is a prime example. What are the implications of the centaur and minotaur models of AI-enabled warfare for logistics during a protracted conflict with China, which analysts caution has key advantages in terms of interior lines, mass, and magazine depth?[59]

Finally, do soldiers trust partnering with AI during battlefield logistics, a practice that experts commonly refer to as human-machine teaming? Further, what are the implications of generational differences across the ranks for trust? Are junior personnel 'digital natives' and senior personnel 'digital troglodytes'? Emerging research provides some answers, finding that soldier trust in human-machine teaming is predicated on a tightly coupled set of instrumental, normative, and operational considerations.[60] However, we need more research to drive AI-enabled capabilities' development, testing, experimentation, and roll-out, especially if U.S. Army senior leaders believe human-machine teaming is integral to future success during great-power war.[61]

---

58 Mark A. Milley, "Strategic Inflection Point," Joint Force Quarterly, Vol. 110, No. 3 (2023), 6-15.
59 Charles Flynn, interview by Thomas Karako, "Strategic Landpower Dialogue: A Conversation with General Charles Flynn," Center for Strategic International Studies. Oct. 12, 2023. https://www.csis.org/analysis/strategic-landpower-dialogue-conversation-general-charles-flynn.
60 Paul Lusheko, "Trust but Verify: U.S. Troops, Artificial Intelligence, and an Uneasy Partnership," Commentary, Brookings. Jan. 22, 2024. https://www.brookings.edu/articles/trust-but-verify-u-s-troops-artificial-intelligence-and-an-uneasy-partnership/.
61 Colin Demarest, "Don't Ditch Soldiers for Machines, Combine Them, Rainey Says," C4ISRNet. Aug. 18, 2023. https://www.c4isrnet.com/artificial-intelligence/2023/08/18/dont-ditch-soldiers-for-machines-combine-them-rainey-says/.

12

DARWIN STRATEGIC BASTION

Mick Ryan

The following is an abbreviated version of an account of the establishment of the Darwin Strategic Bastion (DSB) during the Second Pacific War. It is based on a series of post-war interviews with Captain Bruce McCammon, who served as aide-de-camp to the Australian Commander of Joint Logistics from 2027 to his death in a plane crash in 2029. McCammon, who missed that fateful flight due to a lack of seats on the aircraft, was shortly transferred to an infantry battlegroup and served in the Manus, Rabaul, and Solomons campaigns between 2029 and 2032. Like my previous account of the Battle for Taiwan, I have used the narrative form to portray the crucial role of the strategic bastions in the Second Pacific War.

01 July 2028

Captain Bruce McCammon peered across the still, azure waters at the ochre and green far shore. In Darwin harbor, many naval and commercial ships were moored over the hundreds of square kilometers of this protected anchorage. Stretching from Charles Point in the west to Lee Point in the east, the massive harbor emptied into the Beagle Gulf to the north. And beyond that lay the hostile waters containing a multitude of Chinese crewed and uncrewed underwater vessels, surveilled by the unblinking eye of the People's Liberation Army satellites and very-high-altitude drones.

It had been three weeks since the large-scale Chinese invasion of Taiwan had begun. Leading with cyber, air, and rocket attacks combined with assassination attempts on senior government figures in Taipei and a mass amphibious drone landing, the initial Chinese strike across the Taiwan Strait had experienced mixed results. In the north, an immense Chinese naval force of warships and amphibious vessels had been decisively defeated by a joint American-Japanese air-sea task force. The PLA-Navy losses included one of its aircraft carriers, dozens of destroyers and frigates, and many amphibious and commercial ships. The leadership of the Chinese Communist Party were apparently keeping this a secret from their citizens.

That will not go down well with the Chinese people if, and when, it does get out, mused McCammon. Although, as he then recalled, the CCP had successfully withheld the details of the massacre of protestors in Tiananmen Square for several decades. The deaths of thousands of Chinese citizens in the center of its densely populated capital had been all but erased from open discourse. It would be much easier for the Chinese regime to hide a catastrophe that had occurred far out to sea beyond the view of most of its citizens.

Unfortunately, the Chinese military operations launched towards the southern reaches of Taiwan had experienced more success. McCammon remembered the briefings throughout the middle of June from the intelligence officers embedded in his unit.

The Chinese had successfully landed in southern Taiwan. Masses of uncrewed amphibious robots had killed thousands of Taiwanese soldiers and civilians.

The PLA had landed multiple brigades in the south and were building up for a breakout to the north and east. Huge logistics dumps had been established and were being expanded.

The PLA 71st and 72nd Group Armies were moving steadily north, although suffering significant casualties while doing so.

Chinese sabotage teams intercepted in Sydney and Okinawa.

These reports were in the main theater of operations, and the tranquil waters of the harbor before McCammon could very easily lure one into a false sense of security. Darwin was thousands of kilometers from the landing beaches in Taiwan, which had witnessed enormous casualties on both sides. It was also a long way from the deadly skies where crewed and uncrewed aerial vehicles of every type wrestled in a constant struggle for air superiority.

However, as McCammon well knew, Darwin's physical distance from Taiwan belied just how crucial this location was to the overall war effort waged by a coalition of Taiwan, America, Japan, the Philippines, and Australia. Darwin was one of several strategic bastions that the coalition had established to support combat operations in and around Taiwan over the past several months in the lead up to the anticipated Chinese invasion of Taiwan. It was linked to the conflict by the military support it pushed forward and by its crucial role in receiving the damaged and destroyed bodies of combatant personnel every day. And almost daily, damaged ships limped back into the harbor to await their place on the list of vessels to be repaired at the scarce ship repair berths or alongside the many repair ships moored in the harbor.

These locations, to the south, east, and north of Taiwan, were crucial strategic hubs that functioned as entry points to the western Pacific theatre overseen by the new coalition commander for WestPacific Forces. In these strategic bastions, an array of ship loading and unloading, airfield operations, ship repair, reception and dispatch of battlefield replacements, and the onforwarding of replacement aircraft and equipment took place. No modern combat operation was possible without such massive logistic functions being exercised around the periphery of the established theater of war. The great campaigns of the First Pacific War, first by the Japanese and then by the Americans, had shown just how fundamental logistics were over such an enormous proportion of the earth's surface.

Another characteristic of these strategic bastions was their protection. Despite their lack of proximity to the battlespace, they were still vulnerable to long-range Chinese missiles, drones, cyberattacks, and sabotage operations. As such, each strategic bastion established in the past few months contained multiple air and missile defense units. Everything from the older Patriot air defense system to newer directed energy systems, all linked in a digital air defense network, were present in and around the Darwin region. Strategic cyber defense units worked 24/7 to ensure that all the vital computer systems that kept military logistics and communications running were secure and—above all—trusted by their users.

And on the ground, masses of the new uncrewed ground vehicles undertook a myriad of functions. Some surveilled the many units and establishments in the strategic bastion. Others were combat drones tasked with the close protection of the bastion. But, while there were thousands of such drones in the Darwin area, this number was dwarfed by the many different kinds of uncrewed ground logistic drones.

This time last year, McCammon thought, *I was sitting happily in my Captains' Course at Canungra in southern Queensland. Who would have thought that just a year later I would be here, part of the largest concentration of military logistics since the World War II?*

McCammon knew Darwin's history during the First Pacific War of 1941 to 1945. It had been a major Allied base location in the early 1940s. It was both a significant anchorage and the site of many airfields from which bombers had sortied north to attack the Japanese. Darwin itself had also been bombed on many occasions during that war, as over one hundred

Japanese bombing raids had been conducted in the Darwin area between February 1942 and November 1943. The war cemeteries in places like Adelaide River were a testament to this deadly struggle that had occurred in the skies over Darwin.

It's funny that we now think of that older conflict as the First Pacific War since the beginning of the Chinese invasion of Taiwan. Given everything that has occurred since then, it has certainly earned the title of the Second Pacific War.

McCammon's involvement in the war had begun months before hostilities had broken out. At the beginning of the year, he had commenced a new appointment as the aide to the Commander of Joint Logistics in the Australian Defence Force. While McCammon was an infantry officer and was initially appalled at such an appointment with 'the loggies,' he had quickly gained an appreciation for the hardworking logisticians of the various services with whom his commander associated daily, which became a real fondness for his comrades.

Several months ago, his commander had returned from a compartmented briefing with the Chief of Defence Force in Canberra with news of a very big task ahead. Intelligence assessed a Chinese attack on Taiwan was imminent, and consequently, the long-standing Contingency Plan MAROON FENCE (*who picks those names?*) was to be implemented. Despite the weird name, the plan itself was anything but crazy. It was the outline for one of the several protected logistic bastions that would be established in the western Pacific to support combat operations to protect Taiwan and destroy any Chinese force seeking to attack Taiwan or other locations across the Pacific.

The following weeks had been a whirlwind of activity for McCammon as he was busy supporting his commander in an array of meetings and conferences with Australian and foreign military, state, and local government officials. There were also meetings with a variety of civil companies that would provide many services for the massive new logistic hub.

The first and most important priority had been to secure sufficient real estate in the Darwin area. Fortunately, the Australian military already had large installations in what was known locally as the "Top End." Army, Navy, and Air Force installations all had sufficient room for some expansion. As large as these holdings were, the estimates of the overall footprint on land, on the harbor, and in the air were massive. The navy would essentially have to take over all berthing space at naval and commercial ports in Darwin, and the Darwin airport would be all but closed to civil aircraft so that the surge in military air movements could be addressed. There would also be immense ground transportation requirements. All this necessitated negotiations with the territory government in Darwin and compensation for losses due to the halting of tourism and other revenue losses.

As hard as all that was, these meetings and negotiations had only secured them more room. More dirt. Next was the process of building up accommodation for the expected surge of people, units, and equipment. The initial estimate had been for an increase from the peacetime population of about ten thousand military personnel in the region to at least fifty thousand.

McCammon smiled at that figure. They had busted through that in the second week of operations for the strategic bastion. They were now sitting at around ninety-five thousand and still growing. Logistics is a big business, he mused.

The next order of business was the construction contracts for accommodation and the massive complexes of warehouses, tarmac space, and outdoor storage areas that would be required. This was complicated because Darwin was located in monsoon and cyclone zones, meaning that whatever they built during the dry season in the middle of the year must also be able to survive torrential downpours during the wet season.

He didn't even want to think about what would happen if there was a cyclone!

Once all the issues with accommodation, warehousing, tarmac space, and allocating space to all the logistics, air defense, and ground defense units expected to arrive in a short amount of time, access arrangements had to be made for the allied nations that would be part of this conglomeration.

Fortunately, there was a long-standing status of forces agreement between the US and Australia. The more recent strategic agreements between Japan and Australia also helped. However, these were just a foundation for the more complex discussions around who would be paying for what part of the massive infrastructure expansion in Darwin.

On top of that, other countries that did not have strategic logistics and infrastructure-sharing arrangements were due to arrive. The Brits and Filipinos both expected to deploy significant contingents of their military logistics units to Darwin, as did the Canadians, Chileans, Poles, and Ukrainians.

One long-standing infrastructure irritant for the Americans had been easily dealt with. In 2015, the local government struck a deal with a Chinese commercial entity to lease the Port of Darwin for 99 years. While it had taken a couple of clever pieces of federal legislation to resolve, the Chinese lease had been abrogated, and the management of the port handed over to an Australian company. Problem solved.

McCammon had observed as his commander had participated in the complex legal and financial negotiations that had paved the way for all of these nations to contribute to the construction, maintenance, and protection of what the coalition had agreed would be called the Darwin Strategic Bastion, or DSB.

While all these preparations were taking place, McCammon worked on a separate project for his commander reviewing lessons from contemporary and ongoing conflicts that might inform the design and operation of the DSB. While the war in Ukraine and the recent Ukrainian victory in the war had provided a myriad of insights into contemporary warfare, Chinese aggression against Japan, Taiwan, and Vietnam in the past year had provided valuable insights that would be essential for building a robust and survivable logistics approach for the war they now found themselves in.

Some of the most important lessons involved the evolution of drone warfare. While offensive capabilities had made a major impact in Ukraine and elsewhere, developments in defensive technologies had caught up. This dynamic meant that a situation of parity now existed and, more importantly, that it was possible to establish effective drone defense regimes over large, dispersed logistics complexes like the DSB.

Another lesson was the democratization of digital command and control. While this was crucial for battlefield commanders at all levels, it was also vital for the logisticians who were supporting them. Every level of the supporting logistics enterprise needed up-to-date data on consumption from combat forces of all classes of supply. These could be complemented with advanced algorithms that predicted future consumption rates and informed procurement of all kinds of military materiel.

Finally, the ongoing meshing of military and commercial intelligence collection and assessment was as vital to logistics as it was to combat. Not only could logistics organizations gain better visibility of enemy logistics capabilities, but they could also develop a superior understanding of local, national, and international capabilities that might be able to support military logistics. And, of course, this meshing of intelligence would play a vital role in assessments of threats to the DSB from all the domains.

None of these new technologies, or the development of the DSB, came cheap. One of the responsibilities of the Commander of Joint Logistics was to sit on a national-level committee that managed the financial aspects of the war. After several nano-seconds, The Australian government realized that years of spending only 2% of the national GDP on defense would be manifestly inadequate for the growth in the military and its supporting infrastructure required to support the defense of Taiwan and the wider region from the Chinese. McCammon's boss was one of several defense representatives on this committee and was constantly provided updated estimates of current and anticipated future expenditures on logistics and procurement of equipment and munitions for the military.

Shaking himself from these thoughts, McCammon looked across the harbor waters and the many ships assembled there to the West Arm of the harbor. Located there was the anchorage for ships that carried the many forms of explosive cargoes needed by combat forces in wartime. Ammunition ships and oilers of all types were gathered in this remote part of the harbor, well away from the port and from the natural gas export facility, which still operated from the middle arm of the port.

McCammon also remembers how the planners who had laid out the Darwin anchorage were well aware of the May 1944 West Loch disaster in Pearl Harbor. They were keen to avoid repeating that catastrophe from the First Pacific War, which had killed and wounded hundreds and destroyed precious amphibious ships in the lead-up to the American invasion of the Japanese-held Mariana island chain.

Not only would such a disaster result in a human and operational tragedy, but the forces coalition did not have a single ship to spare in the Pacific. For decades, shipbuilding in the US and Australia had declined, reaching the point that just before the Chinese began their attack on Taiwan, a single Chinese shipyard in Jiangnan alone possessed more capacity than all American and Australian shipyards combined. China's naval shipbuilding capacity in the late 2020s was now over two hundred times larger than that of the United States. Think tanks and congressional reviews into American industrial capacity repeatedly identified this issue in the 2020s. Although well known, very little headway had been made in increasing ship production. As such, there was almost no prospect in the short term of replacing any ship lost during the conflict with China, while the Chinese could probably build dozens of new ships to replace any losses.

Fortunately, the industrial bases in America, Australia, and other partner nations had learned from the war in Ukraine and had rapidly expanded their production of other military materiel such as drones, munitions, advanced strike missiles, and air defense missiles. This military production capacity, which had also contributed to expanding the war stocks of Japan, America, and Australia over the past couple of years, had probably been the difference between quick defeat and holding off the Chinese in the early days of this new conflict.

McCammon's eyes drifted away from the ships laden with munitions to a very different naval facility near the harbor's eastern arm. A large manufacturing facility for uncrewed naval vessels was co-located with the existing commercial port, now run by the Royal Australian Navy. While Australia may not have built many large vessels in the past several decades, it did possess dozens of small boat builders up and down the country's east coast. Last year, a government-private conglomeration had been established to use this small boat-building knowledge, coupled with input from the Ukrainian navy, to begin constructing a large fleet of small- and medium-sized uncrewed naval vessels for surveillance, submarine detection, and strike operations.

The facility in the east arm was pumping out these vessels with a production line that looked like the massive American aircraft production line at Willow Run, Michigan, during the World War II. McCammon and his boss had visited the facility last week. It utilized the latest robotic manufacturing techniques and contained hundreds of 3D printers. There were some humans around, but they were primarily involved in quality control and problem-solving. *Very impressive*, McCammon thought.

For months, these vessels had been conducting operations across the northern approaches to Australia and all the way along the western and eastern coastlines—that was a lot of the earth's surface to cover, McCammon pondered. Doing it with a mass of uncrewed vessels made a lot of sense. Some of these vessels also prowled the entrance of Darwin Harbor and inside the harbor to deter or destroy Chinese attacks on logistic shipping.

Maintaining this fleet was the work of naval and civilian technicians in one of the hundreds of logistics contracts established with civilian companies during the build-up of the DSB. Indeed, this contracting capability was one of the vital elements of the headquarters in which McCammon worked. All kinds of commercial entities were needed to produce materiel; carry goods to and from the DSB; maintain buildings; haul away trash; undertake repair of ships, trucks, and drones; and countless other tasks. The contracting cell was comprised of hundreds of Australian, American, Japanese, British, and Canadian personnel from the military and civilian life working shifts, which allowed contracting to go on twenty-four hours a day, seven days a week, to meet the colossal and ever-evolving requirements of the DSB.

McCammon finally returned his thoughts to the present and the matter at hand. The farewell ceremony was wrapping up with a final, short speech by the commander of the Australian Defence Force. Sitting behind his boss who was also observing the speech, McCammon listened as the senior officer described the role of the Darwin Strategic Bastion and how it had been essential to the deployment of the massive naval task force they were here to farewell.

The largest coalition task force assembled in this war was arrayed across the harbor and slowly making its way into Beagle Gulf. A mixture of crewed and uncrewed vessels, warships and amphibious ships, merchant marine and fast passenger ferries, this task force was deploying to Manus Island and northern Papua New Guinea. For some time, there had been chatter in PLA high command circles about a 'south Pacific endeavor' to draw the Americans and their allies away from Taiwan. To pre-empt such a move, a large coalition force was being dispatched north to occupy these areas and prevent the Chinese from seizing a foothold deeper into the Pacific.

McCammon, who liaised with many of the Australian and coalition military units as part of his regular duties, had heard stories about the many supporting operations that underpinned the deployment of this naval task force. Secret space and cyber operations had been conducted to deceive Chinese satellites surveilling this part of the world. New aerial and maritime drones had been rapidly developed to form a protective bubble around the massive naval task force with its combat and logistics elements while moving north. And there were even rumors of a new, semi-submersible long-range cargo carrier that would provide logistic support between northern Australia and the deployed forces in Papua New Guinea and Manus Island.

The speech ended. McCammon's boss stood up before him and gestured that they would be leaving straight away. While attending these ceremonies was part of being a senior military officer, there were other more important duties for the commander of joint logistics to attend to. Because, while the DSB was now well established, it was likely that coalition forces might need another strategic bastion established for their logistic support once Manus Island was secured.

They had some long days and nights of logistics planning ahead of them.

CONTRIBUTORS

Kevin Benson is a retired professional Soldier. He is a teacher, a consultant and a writer. From July 2002 to July 2003, he was the Director of Plans for Third U.S. Army and the Combined Forces Land Component Command at the beginning of Operation Iraqi Freedom. His experiences in that role are captured in his new book, Expectation of Valor, from Casemate. A 1977 graduate of the United States Military Academy at West Point, his education ranges from military schools such as the School of Advanced Military Studies to the Massachusetts Institute of Technology Security Studies Program. He earned a Ph.D. in American history from the University of Kansas in 2010 and served as a Senator Robert J. Dole Fellow at the Dole Institute of Politics in 2011. He was appointed an Adjunct Scholar at the Modern War Institute at West Point in 2020.

Rich Creed is a retired Army officer who manages the U.S. Army's doctrine program on behalf of U.S. Army Training and Doctrine Command. He has published articles in various Army professional journals and holds master's degrees in strategic and advanced military studies. Rich served at most Army echelons from platoon to theater level and commanded at the company, battalion, and brigade level. He had tours of duty in Germany, Bosnia, Korea, Iraq, and Afghanistan during the course of his 32-year career.

Matt Evers is an active-duty U.S. Marine, logistician, and wargamer, having served over a decade between the Fleet Marine Force and the U.S. Marine Corps Training and Education enterprise. Matt is passionate about connecting innovators, strategists, wargamers, and supply chain professionals to advance national security. He is a graduate of the U.S. Naval Academy and the U.S. Army's Logistics Captains Career Course. He holds a Master of Science in International Transportation Management from SUNY Maritime College and a Master of Military Studies from the U.S. Marine Corps Command and Staff College.

Tim Gilhool is the Command Historian for the U.S. Army Combined Arms Support Command and Fort Gregg-Adams, Virginia. He is a retired Logistics Corps officer who served in Germany, Korea, and multiple CONUS assignments, as well as combat and disaster relief deployments to El Salvador, Iraq, Afghanistan, and New Orleans. He served 27 months as the battalion commander for the 782nd Brigade Support Battalion, 4th Brigade Combat Team, 82nd Airborne Division at then-Fort Bragg, NC and Kandahar, Afghanistan. He received a B.A. from the University of Michigan, an M.A. from the University of Richmond, and is an alumnus of the U.S. Army School of Advanced Military Studies. He is also a contributor to Army University Films France '44: The Red Ball Express and D-Day: Planning the Impossible.

Ron Granieri is an Associate Professor of History in the Department of National Security and Strategy at the United States Army War College, as well as a Templeton Education Fellow at the Foreign Policy Research Institute, where he is Director of the Center for the Study of America and the West. A graduate of Harvard and the University of Chicago, he is a specialist in German History, European-American Relations, the Cold War, and contemporary politics and co-editor of The Bondian Cold War: Global Connections of a Cold War Icon.

Jim Greer is an Associate Professor at the U.S. Army School of Advanced Military Studies and holds doctorate and master's degrees in education, a Master of Science degree in National Security Studies, and a Master of Military Art and Science degree focused on Military Strategy. In addition to his passion for educating future leaders, he is a futurist, avid science fiction reader, and wargamer. He is a retired U.S. Army colonel who served in Cavalry and Armor positions from platoon through brigade command, serving overseas in Europe, the Middle East, and Asia.

Richard Killblane served as an infantry and Special Forces officer and is a veteran of Panama (Operation Just Cause) and the Central American Counterinsurgency. He graduated from the United States Military Academy at West Point in 1979 and earned his M.A. in History from the University of San Diego in 1992. He retired as the U.S. Army Transportation Corps Historian with several deployments to Iraq and Afghanistan. He has authored a number of articles and books on military logistics, and his recent book, Delivering Victory, is the culmination of 19 years of research and observation as an Army logistics historian.

Jon Klug is the Associate Dean at the United States Army War College. He is Associate Professor and Admiral William F. Halsey Chair of Naval

Studies. Jon is a veteran of multiple deployments, an experienced staff officer in American and NATO headquarters, and an award-winning military history instructor at the Air Force Academy, the Naval Academy, and the Army War College. He holds a Ph.D. in History from the University of New Brunswick, Canada. His dissertation was titled "Building a Global Navy: U.S. Naval Logistics, 1775-1941."

Steve Leonard is an award-winning faculty member at the University of Kansas School of Business, former senior military strategist, and career writer and speaker with a passion for developing and mentoring the next generation of thought leaders. Published extensively, he pens a weekly editorial column, Point of Departure, where his writing focuses on issues of leadership and leader development. He is a member of the editorial review boards of Military Strategy Magazine and the Arthur D. Simons Center's Interagency Journal; and the author, co-author, or editor of several books, including Power Up (2023), To Boldly Go (2021), Why We Write (2019), Winning Westeros (2019), and Strategy Strikes Back (2018).

Paul Lushenko is a U.S. Army Lieutenant Colonel and an Assistant Professor and Director of Special Operations at the U.S. Army War College. He is also a professorial lecturer at George Washington University's Elliott School of International Affairs, a Council on Foreign Relations term member, a Senior Fellow at Cornell University's Tech Policy Institute and Institute of Politics and Global Affairs, and a Non-Resident Expert at RegulatingAI. His work lies at the intersection of emerging technologies, politics, and national security, and he also researches the implications of great power competition for regional and global order-building. Paul is the author and editor of three books, including Drones and Global Order: Implications of Remote Warfare for International Society (2022), The Legitimacy of Drone Warfare: Evaluating Public Perceptions (2024), and Afghanistan and International Relations (under contract). Paul has written extensively on emerging technologies and war, publishing in academic journals, policy journals, and media outlets such as Security Studies, Foreign Affairs, and The Washington Post. He earned his Ph.D. and M.A. in International Relations from Cornell University, holds an M.A. in Defense and Strategic Studies from the U.S. Naval War College, an M.A. in International Relations and a Master of Diplomacy from The Australian National University, and a B.S. from the U.S. Military Academy.

Francis J. H. Park recently retired after 30 years in the U.S. Army as an armored cavalryman and strategist. His final assignment before retirement was Director, Basic Strategic Art Program at the U.S. Army War College. As a strategist, he has served in planning, strategy and policy assignments at virtually every echelon from division to national military staff, including deployments to Kuwait, Iraq, Afghanistan, and Qatar. As a historian, he has served at the U.S. Army Center of Military History and the Joint History Office. His body of work includes authorship of the 2014 Army Strategic Planning Guidance, 2018 National Military Strategy, and Joint Doctrine Note 2-19, Strategy. He oversaw the 2017 rewrite of the Joint Strategic Planning System and three revisions to the Unified Command Plan, and authored parts of the U.S. Army's official history of Operation Enduring Freedom. A graduate of the Johns Hopkins University, St. Mary's University of San Antonio, Texas, and the U.S. Army School of Advanced Military Studies, he holds a PhD in history from the University of Kansas.

Gus Perna co-led the U.S. government's COVID-19 response in 2020 with Operation Warp Speed and served as head of the U.S. Army Materiel Command, where he managed redistribution of critical munitions to the Pacific and fixed the Army's housing crisis. One of the military's most skilled and most senior logisticians, General Perna worked to build manufacturing capacity and the end-to-end supply chain needed to get vaccines distributed "at warp speed." Leveraging public-private partnerships, he accelerated the operation beyond what anyone thought possible—all while the vaccines were still in development. General Perna retired from the U.S. Army in August of 2021 after 38 years of distinguished service in which he was awarded the Defense Distinguished Service Medal, Distinguished Service Medal with three Oak Leaf Clusters, Defense Superior Service Medal with Oak Leaf Cluster, Legion of Merit, and Bronze Star Medal with Oak Leaf Cluster. In 2021, the Association of Business Executives for National Security honored him with the Eisenhower Award.

Michael Posey is a U.S. Navy Commander and an Assistant Professor and Director of Maritime Operations at the U.S. Army War College. An active-duty naval flight officer with a subspecialty in Information Systems and Operations, he holds a Master of Strategic Studies from the U.S. Army War College, a Master of Military Operational Art and Science from the Air Command and Staff College, a Master of Business Administration from the University of Florida, and a B.S. from Carnegie Mellon University.

Mick Ryan is a strategist and former Australian Army senior leader and had the honor of commanding soldiers at troop, squadron, regiment, task force and brigade levels before retiring in 2022. He has a long-standing interest in military history and strategy, advanced technologies, organizational innovation, and adaptation theory. Mick is the author of three books: War Transformed (2022), White Sun War: The Campaign for Taiwan (2023) and The War for Ukraine (2024). Mick is currently the inaugural Senior Fellow for Military Studies at the Lowy Institute in Sydney, and an adjunct fellow at the Center for Strategic and International Studies in Washington DC.

Sydney Anne Smith is a retired U.S. Army Logistics Corps officer who served 28 years on active duty, commanding at the battalion and brigade level, as well as serving as the Assistant Chief of Staff, G-4 for the 82nd Airborne Division at Fort Bragg (now Fort Liberty) NC, and the CJ4, RC-South, Kandahar Afghanistan. She has also deployed to support disaster relief in El Salvador and served two combat tours in Iraq. Since retirement, she has been a member of the Senior Executive Service, with assignments as the President, Army Sustainment University and Director, Army Supply Policy (G-44S), Headquarters, Department of the Army. She is a graduate of Davidson College, the U.S. Army School of Advanced Military Studies, and the Eisenhower School for National Security and Resource Strategy.

Stacy Tomic is a U.S. Army Colonel and a Faculty Instructor and Director of Theater Logistics at the U.S. Army War College. An active-duty logistics officer, she holds a Master of Strategic Studies from the U.S. Army War College, a Master of Business Administration from the University of Tennessee, and a B.S. from the United States Military Academy at West Point.

Joe Walden is an Associate Teaching Professor for Supply Chain Management at the University of Kansas School of Business, where doctoral research focused on developing supply chain management curriculum. A former U.S. Army logistics officer, he served as the Director of Distribution Management in Kuwait and commanded the National Training Center Theater Support Command. An alumnus of the Army's School of Advanced Military studies, he is the author of three books on supply chain management and leadership, six textbooks on supply chain management, and the editor of three other textbooks on logistics. He is a Certified Fellow in

Planning and Inventory Management for the Association for Supply Chain Management. He has also taught procurement for Webster University and operations management for Haskell Indian Nations University.

Index

A
Afghan Army   84, 117
Afghanistan   2, 22, 30, 34, 83–84, 87, 102, 112–30, 180–81, 184, 195–96, 198
Afghan National Army (ANA)   84, 116–17
Afghan National Defense   114, 117
Afghan National Defense and Security Forces. See ANDSF
Afghan National Security Forces (ANSF)   115
Afghan Surge   120–23
AI-enabled logistics   177, 184
AI-enabled warfare   171–72, 177
  centaur and minotaur models of   177, 185
  patterns of   173–74, 177
AI-enhanced capabilities   169–70
air and seaports   6, 57, 106, 108, 111
aircraft, commercial   65
airlift   49, 53–56, 84, 126
Alexander   8–12, 16–18, 21–22, 24
Alexander's strategy   21–22
Allied Joint Doctrine for Logistics   174–76
allies   26, 35, 37, 39, 50, 52–55, 61, 70, 74, 82, 115, 130, 160, 165, 182, 194
Alternate Afghan Supply Routes   122

American defense industries   37, 46
American industry   38–39, 41–43, 45, 48
ammunition   38, 43, 71, 78–81, 83, 91, 96, 101, 134, 136–37, 140, 177
ANA (Afghan National Army)   84, 116–17
ANDSF (Afghan National Defense and Security Forces)   114, 117, 124, 127
ANSF (Afghan National Security Forces)   115
APS (Army Prepositioned Stock)   6
Army Logistician   160, 177, 179
Army Materiel Command   84, 126, 198
Army Prepositioned Stock (APS)   6
Army Sustainment   121, 147, 160–61
Army War College   112, 197–99
arsenal of democracy   2, 35–49
Artifical Intelligence and Logistics   171–85
Artificial Intelligence   3, 156, 169, 171–72, 178, 184–85
art of logistics   2, 5, 8
art of war   18, 20–21, 152, 165
ASCM (Association for Supply Chain Management)   19–20, 147, 156, 158–59

Association for Supply Chain Management. See ASCM
Australian Defence Force   189, 194

B
Baghdad International Airport (BIAP)   80–81
bases, operational   139
battle captains, great   2, 7, 9, 11, 16, 18
Berlin   49–50, 52–56, 77
Berlin Airlift   2, 49, 54–56
BIAP (Baghdad International Airport)   80–81
bridges   74–75, 81, 140, 154

C
campaigning   3, 7–8, 11, 112–13, 127, 164
campaign planning   9, 185
campaigns   2–3, 7, 10–14, 16–17, 49, 57, 59, 72, 74, 77, 84, 104, 106, 110–11, 113–14, 120–21, 127–28, 131–32, 135, 137, 139, 145, 199
  conduct of   57, 130
  great   2–3, 188
capabilities, industrial   47–48
capacity   35–36, 116, 124, 133, 168, 175, 192
capital, human   153, 157–58, 164
cargo   59, 67, 69–73, 78, 80, 103, 120, 122, 125, 128
  offload   69–70
cargoes   119, 121–22, 126, 128
cargo ships   62, 106–7
CDC (Corps Distribution Center)   80
centaur   172, 177, 181, 185
centaur warfare   172, 178–79
centaur warfighting   172

CENTCOM (Central Command)   68, 101–2, 104–5, 107, 112, 179
CENTCOM planners   105–6
Central Command. See CENTCOM
CFLCC (Combined Forces Land Component Command)   100–101, 103–5, 107–9, 195
CFLCC planners   105, 107
changing character   142, 152, 155, 168
China   73, 77, 131, 150, 162, 184–85, 192
Clausewitz   11, 15–16, 152, 178
coalition   28, 79, 109, 112–14, 116–17, 121, 124, 127–29, 131, 187, 191
coalition forces   79, 108–9, 114, 120, 124–25, 127, 180, 194
combat operations   112, 124–25, 135
combat power   3, 6, 58, 63, 68, 71, 84, 118, 135, 143
combat service support (CSS)   97, 103, 109, 150, 160–61
combat support (CS)   97, 103, 105, 118, 154
Combined Forces Land Component Command. See CFLCC
command, combatant   58, 65, 123, 185
commander of joint logistics   189, 191, 194
communication
  air lines of   76–77, 100, 111
  ground line of   76–77, 114, 180
complex adaptive systems   14, 147, 155, 157, 159, 161, 163–64, 166

Contested Environments 148, 151
contested logistics 110, 144–45, 147–49, 151–53, 160, 165, 168
contested logistics environment 147, 152
convoy security 78–79
coordinate 37, 59, 66, 108
coordination 100, 108, 110, 112, 124, 128, 176
Corps, Marine 60, 150, 167
Corps Support Command (COSCOM) 80–81
COSCOM (Corps Support Command) 80–81
CS. See combat support
CSS. See combat service support
culmination 3, 10, 113, 121, 129, 132, 136, 140, 142, 144, 196

D
Darwin 187–90
Darwin Strategic Bastion. See DSB
Defense Industrial Base. See DIB
Defense Logistics Agency (DLA) 122, 126
Department of Defense (DoD) 41, 46–48, 58, 63, 65–66, 124, 138, 148, 160, 165–67, 184
deploy 7, 58, 60–61, 63–68, 83, 91, 133, 190
deploying 58, 102, 115–16, 194
deployment 58, 60–61, 63–68, 71, 95–96, 104, 108, 113–14, 121, 174, 194, 196, 198
deployment planning 6, 27, 66, 96
deploy troops 59, 65
design, operational 132, 144
DIB (Defense Industrial Base) 46–48, 150
distances, operational 133, 144

distribution operations 33, 80
DLA (Defense Logistics Agency) 122, 126
doctrine
   joint 160, 163, 174–75
   service-oriented 175–77
DoD. See Department of Defense
drones 142–43, 169, 188, 192–93
DSB (Darwin Strategic Bastion) 3, 186–87, 189, 191, 193–94

E
economy 9–10, 37, 156, 158, 161–62, 176–77, 181, 183
   national 45, 128, 154
efficiency 10, 66, 71, 151, 157, 169, 176, 181, 183
environment 20, 22, 33, 98, 116, 147, 149, 154–55, 159, 162, 166, 179
   external 13–14, 17
example of logistics driving strategy 26–27, 32–34
examples of logistics driving strategy and strategy driving logistics 32, 34

F
force deployment 104, 112–13
force flow 102–6, 127
force packages 102–3, 106, 108
forces
   deploying 109, 124
   movement and support of 174
   special 117, 196
   special operations 102, 111, 116–17, 125
Forces Command (FORSCOM) 103, 107
force sustainment 113, 123

FORSCOM (Forces Command)   103, 107
fuel   49, 71–72, 75, 78–81, 89, 91, 96, 101, 103, 177, 179–80
fuelers   89–90, 92, 95
fuel supply   75, 78, 182
fuel transfer pump   89–90
functions   20, 41, 57, 74, 83, 98, 147, 149, 155–57, 159–60, 168–69
 joint warfighting   160, 167
functions of logistics   57, 74, 154, 156

G
Garcia, Diego   67–68
Genghis Khan   9–10, 12
global positioning system (GPS)   67
GPS (global positioning system)   67
great captains   8, 12–14, 17–18
great captains of battle   9, 12
ground transportation   64

H
harbor   187–89, 192–94
human-machine teaming   185
human oversight   170–73, 184

I
implications   28, 132, 143, 169–70, 172, 174, 177, 184–85, 197
improvisation   9, 11, 176, 179–81
industrial base   2, 50, 100, 192
industry   36–37, 40–41, 44–45, 47–48, 148–49, 158, 161, 165
 commercial   20, 30, 34
information environment   149, 155–56, 163–66
International Security Assistance Force. See ISAF
intuition   14

Iraqi Army   79
ISAF (International Security Assistance Force)   83, 116–18, 120, 125
ISAF Joint Command (IJC)   120

J
Joint Chiefs of Staff   66, 100, 113–14, 121, 150, 160, 163, 165, 174, 182
Joint Logistics   114, 150, 160, 174–77, 186, 189, 191, 194
Joint Operating, Planning System (JOPS)   66
Joint Operational Warfare   151, 165
Joint Operations, Planning, and Execution System (JOPES)   66
Joint Staff   104–5, 113, 132
Jomini   11, 152–54
JOPES (Joint Operations, Planning, and Execution System)   66

K
KBR (Kellog, Brown, and Root)   81
Kellog, Brown, and Root (KBR)   81

L
landing craft   70–71, 76, 78
large scale combat operations   130
large-scale combat operations   35, 79–80, 130, 182
Large-Scale Combat Operations. See LSCO
large-scale invasion of Ukraine   130–31
Learning, machine   178, 183

learning culture 157–58
levels
 strategic 100, 111, 131, 169–70
 tactical 131, 151, 166–67, 169, 177
limitations 70, 127, 183
LOCs 118–19, 125, 127, 129
logistical 24, 54, 57, 87, 90, 95, 103–4, 111, 118
logistical bases 118, 135, 139
logistical capacity 3, 127
logistical constraints 21, 151
logistical footprint 80–81
logistical hubs 116, 128, 137, 139
Logistical Literacy 2, 89–99
logistical management 49, 55
Logistical Operations 2, 57, 59–85, 174, 177
logistical planners 108, 183
logistical planning 19, 22
logistical requirements 108, 120, 184
logistical support 26, 76, 90, 121, 133, 135
logisticians 2–3, 6, 19, 21–22, 27, 32, 34, 71, 73, 83–85, 87, 96–97, 106, 109, 113, 125–26, 155, 176, 178, 181–83, 185, 191, 195
 senior 79, 104, 198
logistics 2–3, 7–9, 16–25, 27–34, 49, 55–57, 70–72, 87–93, 95–101, 103–5, 111–13, 121–23, 127–45, 147–61, 168–71, 173–79, 181–85, 190–92
 battlefield 170, 177, 184–85
 just-in-time 21–22
 operational 119, 128, 151
 strategic 120, 190
 study 72, 111
 tactical 87, 92, 128, 183
logistics activities 33, 151, 177
logistics and strategy 2–3, 9, 33

logistics and sustainment 98, 176
logistics and sustainment operations 174–75
logistics capabilities 23, 25, 27, 179
logistics clusters 25, 33–34
logistics command 78, 82, 179
logistics constraints 26–27, 31–32
logistics corridor 28
Logistics Drive Strategy 2, 20–35
logistics driving strategy 19, 25–27, 32–34
logistics forces 102, 104, 106
Logistics Management 156–57, 159
logistics nodes 137, 141
Logistics Officer 90, 183
logistics on strategy 20, 23
logistics operations 7, 29, 57–58, 82, 85
logistics packages 89, 91
logistics planners 5, 26, 104, 179–82
logistics planning 27, 194
logistics principles of responsiveness and improvisation 179, 181
logistics strategy 29, 33
logistics supplies 22–23, 31
logistics support 22, 26, 31, 78, 140
logistics system 10, 30
logistics units 101, 103
LSCO (Large-Scale Combat Operations) 35, 79–80, 130, 132, 182–83

M
maintenance 19, 47, 57, 71, 73, 88–90, 93–94, 96, 159, 174–75, 183

major operations 111, 130–32, 135, 144–45
maneuver, operational 113, 182
maneuver combat forces 121, 124
maneuver units 57, 81, 84, 91, 96
materials 21, 31, 41, 44, 47, 126, 133, 151, 156–58, 160–61, 168
military campaign 112–13
Military Concepts 151, 154
military doctrine 152–54
Military History 20, 23, 26–27, 42, 76, 115–16, 121, 198–99
military logisticians 9, 157
military logistics 20, 148, 150–51, 154–57, 161, 188, 196
military materiel 191–92
military objectives 58, 170
military operations 23, 31, 57–58, 64, 84, 156, 171, 173–76
military power 112, 164, 167
military requirements 46–47
military strategy 2–3, 19–20, 25, 112, 152, 196
military strategy and logistics 2–3
military theorists 9, 20
minotaur 172, 177
minotaur model 177, 181–83, 185
minotaur warfare 172, 174, 181, 183
movement, strategic 22–23, 113
munitions 37, 43, 47–48, 133, 135–36, 138, 141, 150, 192–93

N
national defense 46, 153–54
national security 195, 197–98
national support element. See NSEs

NATO (North Atlantic Treaty Organization) 7, 112, 117–19, 126, 133, 174
NATO in Afghanistan 117, 119
NATO nations 133, 138, 141
NATO's Redeployment 125–26
naval logistics 151, 153, 197
Naval Overseas Transportation Service (NOTS) 61
Navigation System with Timing and Ranging (NAVSTAR) 67
NAVSTAR (Navigation System with Timing and Ranging) 67
navy 23, 41, 44, 61–62, 67, 70, 151, 153, 189
NDN (Northern Distribution Network) 122–23, 125–26, 128, 180
NDN North 122
NDN routes 125
North Atlantic Treaty Organization. See NATO
Northern Distribution Network. See NDN
NOTS (Naval Overseas Transportation Service) 61
NSEs (national support element) 117–18, 127

O
OEF (Operations Enduring Freedom) 65, 114
offensive 67, 75, 136–39, 141–42
offensive operations 31, 69, 140, 143
OIF. See Operation Iraqi Freedom
Operational 3, 130, 132, 135
operational approaches 124
operational areas 22, 78, 121
operational art 2, 57–58, 98, 100, 112, 131, 142, 155–56, 167

operational effects   134–35, 138, 143–44
operational environments   152, 155, 163, 165, 167–68
operational impact   139
operational level   100, 110, 131, 140, 142, 183
operational level of war   100, 108, 111, 170, 180
Operational Level of War and Logistics   2, 101–11
operational performance   157–58
operational performance in supply chain management   157–58
operational warfare   2, 130–31, 133–45
operation art   131–32
Operation Desert Shield   30, 65, 91, 179
Operation Desert Storm   5, 46, 67, 79
Operation Enduring Freedom   65, 114, 198
Operation Iraqi Freedom (OIF)   7, 19, 27, 29–30, 32, 65, 67–68, 79, 81, 195
operations
  amphibious   27, 63, 76
  defensive   67, 132
  littoral   76
  multidomain   148, 160
  multi-domain   132, 166
Operations Enduring Freedom (OEF)   65, 114
operations management   30, 200

P
PakGLOC (Pakistan Ground Line of Communication)   84
Pakistan GLOC   121–22, 125–26
Pakistan Ground Line of Communication (PakGLOC)   84
Panama Canal   25–26, 33, 52
pauses, operational   27, 57, 75, 80
personnel   19, 23, 27, 31, 64–65, 67, 70, 82, 123, 133, 147, 151, 171, 174–75
personnel services   160, 175
Phase IV   101, 103, 108
phases   82, 132, 138–40, 144–45
physical domains of warfare   149, 155, 166
planners   32, 71, 95, 97, 100–102, 105–6, 108–9, 171, 183, 185, 192
planning   8–9, 18, 26–28, 44, 61–63, 66, 87, 90, 96–97, 100–101, 104, 106, 108, 112–13, 121–22, 128, 142, 159, 174–76, 196, 198, 200
plans   5, 25, 30, 34, 51–52, 54–55, 65–66, 82, 98, 101, 106–8, 110–12, 125, 172, 176, 189, 195
  strategic   26–28, 100
policymakers   100–102, 169
POMCUS (Prepositioning of Materiel Configured in Unit Sets)   67
ports   22, 26, 28, 59–60, 64, 68–70, 72–77, 84, 102, 107, 111, 190–92
  coastal   23–24
ports of debarkations   69
ports of embarkation   59, 61
principles of logistics   9, 170, 174, 176–79, 181
principles of sustainment   7, 17, 176
Principles of Sustainment and Logistics in Alexander's Shadow   122, 181
protection forces   110
provisioning   8–9, 16

Q
Quartermaster Corps   23, 73, 80

R
rail   23, 25–26, 33, 59–61, 64, 67–68, 72–73, 76, 132, 136, 138
railhead   73, 75, 136
rail hub   25
rail networks   24
RAND Corporation   31, 151, 156
redeploy   82–83
redeployment   82, 108
requirements, operational   176
reserve component units   103–4
retrograde   30, 58, 82–85, 108–9, 120, 124–28
retrograding   83–84, 125, 128
reverse logistics   30–31
reverse logistics strategy   30
RFFs   102–3, 105
Royal United Services Institute   132, 137–38, 141–42
RSO&I   57, 69, 71–72, 84
RSO&I and strategic deployment in reverse   58, 83
Russia   2, 122, 130–37, 139, 142–43, 181, 184
Russian Army   137, 140–41
Russian attacks   133–34, 136, 142
Russian forces   130, 135–38
Russians   130–31, 133, 135–43
Russian SMO   135, 140, 142, 144
Russia's War in Ukraine   143, 169
Russo-Ukraine War   47–48, 142, 144

S
science of logistics   17, 131
science of operations   131
seaports   6, 57, 60, 70, 74, 106, 108, 111, 113, 115, 122, 126, 132
Second Pacific War   186, 189
security force assistance   120–21, 124
sequencing   102, 105–9, 111
services   57, 60, 71, 84, 158–59, 161, 168, 174–75, 189, 198
set the theater   110
shipping   33, 60, 126
ships   33, 39–41, 48, 60, 62, 64, 67–68, 70, 73, 77, 82, 102, 105–7, 126, 192–93
shortage   23, 30, 74–75, 94
simplicity   9–10, 176, 181, 183
SOLOC (Southern Line of Communication)   75
Southern Line of Communication (SOLOC)   75
staff   66, 90, 96, 100, 104–6, 111, 113–14, 121, 150, 160, 163, 165, 174, 182, 199
staff officers   26, 97, 197
strategic bastions   186–88, 190, 194
strategic deployment   57–61, 64, 67, 69, 83, 144
strategic intuition   15–16
strategic objectives   58, 100, 131
strategic planning   8, 11, 22–23, 25, 27
strategic thinking   11–14
strategists   2–3, 27, 32, 34, 195, 198–99
strategy
 campaign   25, 27, 120
 coalition's   113, 127
strategy and logistics   2, 19–21, 24, 34
strategy and tactics   22, 131, 154
Strategy Drive Logistics   2, 20–35
strategy driving logistics   19, 26–28, 32, 34
strategy of movement   21
study of logistics   85, 145

Sun Tzu 20–21, 29, 34
supplies 16, 19–24, 26–31, 35, 40, 44–45, 48–49, 53, 57–58, 61, 63, 67, 69–75, 77–78, 80–84, 89, 136, 138, 140–41, 149, 151–52, 155, 157–59, 174–75, 180, 191
  medical 71, 183
  move 28–29
  non-lethal 121–22
  tons of 54
supply base 77
supply chain activities 153
supply chain environment 2, 147–69
supply chain environment theory 152, 166
supply chain management 19, 34, 147, 155–61, 199
supply chain management discipline 148, 155
supply chain management theories 152, 155–56
supply chain operations 156–57
Supply Chain Operations and Reference (SCOR) 156
supply chain products 158
supply chain resilience 149, 156
supply chains 3, 45, 138, 148, 155–64, 166–68, 179, 200
  digital 156
  larger 147
  nature of 155, 157
Supplying Washington's Army 22–23
supply lines 8–10, 33, 57, 78, 136, 140, 142, 182
supply routes 79, 83, 110, 114, 180
supply systems 83, 87
support 22, 24–27, 30–31, 33–34, 37, 53, 58, 61–62, 78, 81, 87, 91–94, 102, 104, 107, 109, 116–17, 120–21, 123, 127, 131–33, 135, 150, 160–61, 170–71, 174–76, 180, 182–83, 192
  combat service 97, 103, 160
  logistic 27, 194
support combat operations 187, 189
Support Command 79, 179
support structure 26, 101
sustain 7–8, 10, 14, 35, 53, 58, 67, 71, 74–77, 80, 98, 100, 102–5, 109–10, 112, 116, 118, 123, 125, 128, 152
sustainability 156, 175
sustain battles 100, 108
sustaining 7, 9, 16–17, 63, 72, 109, 111, 118–19, 153
Sustaining Strategy 2, 6–98, 102, 104, 106, 108, 110, 114–204
sustainment 7–9, 57–58, 72, 76–77, 80, 97–98, 110–15, 117, 121–22, 128, 160–61, 173–76, 178–79, 181–82
sustainment forces 110, 121, 127
sustainment operations 77, 80, 84, 169–70, 174–76
systems 13, 21, 48, 51, 66, 88, 93, 110–11, 134, 155–59, 161, 163, 171
systems thinking 12–13

T
tactical level of war 170, 177, 180–81
terms logistics 147, 159–60
theater 31, 58, 61–62, 66–67, 69, 72, 76, 80, 82, 103–4, 108–11, 113, 115–16, 121, 123–28
  austere 115, 127–29
theater commander 58, 72
theater commands 83

Theater Distribution Center   32, 80
theater level   113, 195
theater logistics   72, 80, 123, 179, 199
theater logistics hubs   126, 128
theater logistics infrastructure   123, 126
theater of operations   57–58, 61, 82, 113–14, 116, 123, 126, 128, 132, 170, 187
theater of war   100, 102, 111, 113
theory, game   12–13
throughput capacity   6
time phased force deployment list. See TPFDL
Time Phase Force Deployment Data (TPFDD)   66
TPE (theater-provided equipment)   125
TPFDD (Time Phase Force Deployment Data)   66
TPFDL (time phased force deployment list)   32, 102–3
TRANSCOM (Transportation Command)   65, 102, 104–5, 122, 150, 166
transit   26, 33, 68, 128, 144
transition   66, 114, 124, 126, 132, 136
transportation
  modes of   84
  strategic   58, 82
Transportation Command. See TRANSCOM
transport fleet   63
transports   21–22, 27, 59–63, 73, 82, 102, 126, 174, 179
troop-contributing nations   117–18, 127–28

U
UAF (Ukrainian Armed Forces)   131, 133–34, 138–43
UK   137–38, 141–42
Ukraine   2, 130–45, 150, 162, 169, 191–92, 199
Ukraine logistics   131–32
Ukrainian Armed Forces. See UAF
Ukrainian drones   143
understanding of logistics   25, 98
United Kingdom   53, 151
units
  deployed   61
  deploying   102
  supporting   61, 65
Unmanned War   170–71
USTRANSCOM (U.S. Transportation Command)   65, 102, 104, 122–23, 126, 166

V
Vegetius   8–9
Vego, Milan   151, 165
vehicles   36, 45, 71, 78, 85, 90, 92–94, 98, 110, 125, 136

W
war, next   61, 63, 73, 77, 148
warfare   3, 7, 9–14, 23, 110–12, 130, 149, 154–55, 165–66, 169, 172, 174
  ancient   7, 9
  contemporary   182, 191
  drone   143, 191, 197
  expeditionary   7, 113
  modern   18, 134
war in Afghanistan   2, 129
war in Ukraine   2, 130–33, 135–45, 191–92
WIB (War Industries Board)   37–38
WPB (War Production Board)   41, 48

www.ingramcontent.com/pod-product-compliance
Lightning Source LLC
Chambersburg PA
CBHW041216130526
44590CB00062BA/4263